Creating Pictures with Preserved Flowers

Creating Pictures with Preserved Flowers

Maureen Foster

Drawings by Bryan Foster
Photography by Ron Brockland, LRPS, and Bryan Foster
Pictures and plaques designed and made by the author unless otherwise stated

Pelham Books

First published in Great Britain by
PELHAM BOOKS LTD
52 Bedford Square, London WC1B 3EF
1977

ISBN 0 7207 0984 9

Filmset and printed in Great Britain by
BAS Printers Limited, Over Wallop, Hampshire
and bound by Dorstel Press Ltd, Harlow

In memory of my father, who many years ago designed and made a suitable frame for my very first preserved flower picture. Without his initial help I might not have had this opportunity of helping others to design and make preserved flower pictures.

Contents

List of illustrations

Line drawings *page*

Acknowledgements

I would like to extend my grateful thanks to the following people, for it is their help and co-operation that has made it possible for me to cover a far wider aspect of preserved flower pictures than would have otherwise been possible: to Mr Donald McNarry for his photograph of a picture that he commissioned me to make (Plate 00); to Miss Phylis Edwards, Botany Librarian of the Natural History Museum, for her valuable help in research; and to the following people for allowing me to include their work in this book – Miss Molland for Plate 14; Mrs Bagwell for Plate 20; Miss Broughton for Plate 22; Mrs Sully for Plate 25; Mrs Harris for Plate 27; and last, but by no means least, to my husband Bryan whose help extended far beyond that of artist and photographer – a very special thank you.

Introduction

In my first book, *Preserved Flowers – Practical Methods and Creative Uses,* I included a small section on making plaques and pictures. Since then I have designed and made many more, and this book is devoted entirely to this fascinating aspect of the art of preserved flowers, which is becoming an increasingly popular and much practised form of flower arranging, particularly among flower arrangement societies.

Designing and creating plaques and pictures with preserved plant materials can be enjoyed by people of all ages. In my recently published book for children, *Flower Preserving for Beginners,* you can see plaques made by children from seven to nine years old. A picture made by an old lady of eighty-six is reproduced on page 63 of this book.

Creating three-dimensional pictures with preserved flowers is not only great fun, but can be an extremely interesting and rewarding hobby – and above all inexpensive! Only simple equipment and materials are needed, most of which can be found in and around the home. Flowers, too, need not be exotic – most of those I have preserved to make the pictures illustrated are simple flowers from the garden and hedgerow.

A picture is primarily something which gives us pleasure and harmonizes with the decor of the room in which it hangs. A flower picture can be created from preserved flowers to do just that, no matter what our taste in colour or furnishing, and, what is more important, we need not be born artists to achieve good results. You may ask, 'What is a preserved flower?'. First and foremost, it should not be confused with a pressed flower. In recent years the art of pressing flowers has been revived from the Victorian era, but although pressed flowers are used to create beautiful designs

it is not always possible to retain the rich colourings of fresh flowers, especially the pink and red shades, and it is often necessary to dissect the flower before it can be pressed perfectly flat. Nature has given flowers wonderful colours and contours. Years of experimentation have taught me how to preserve flowers with chemicals and still retain these unique qualities. While there is no real substitute for fresh flowers, with a little practice you will be able to preserve flowers so successfully that they are practically indistinguishable from fresh ones.

It is very important that flowers for preserving should be really fresh, and so I have only used flowers that are found in Britain. As most of these flowers also grow in other countries and in particular the United States of America (many are in fact natives of other countries), the fun of creating pictures with preserved flowers can be enjoyed by those living on both sides of the Atlantic.

For convenience, I have used the word 'flower' as a collective term to include the many other forms of plant materials used in my pictures, such as grasses, cones and seedheads. Many of the latter are equally as useful as flowers because of the interesting form or texture they provide, and unlike flowers they can be collected by the tourist abroad and successfully transported home. Plate 5 shows the additional interest these materials can provide.

As you turn the pages of this book you will find photographs of many different types of plaques and pictures. While designing and making this collection I tried to think of the many types of settings in which preserved flower pictures could be used, and also the different occasions. As a

PLATE 1 TOOLS AND MATERIALS The very simple tools and materials that are used to create pictures with preserved flowers: ruler, small pointed scissors, pencil, floral scissors, scissors for cutting fabric, Gold Finger paint, a selection of artist's paintbrushes, artist's oil paints, Bostik or clear adhesive glue, Surform tool, Clam PVA glue, shoe polish, modelling clay, white spirit, Stanley knife, chalk, picture hooks, sticky hanging tabs for plaques, spreader for glue.

result, I hope my book will contain inspiration for all who use it. Attractive and inexpensive frames can be easily made from cardboard, following the step-by-step illustrations. Restoring antique frames and adapting them to take three-dimensional preserved pictures is also made easy by step-by-step instructions and illustrations, and those who wish to make something completely different can use my latest idea of converting old clocks and watches into frames for preserved flower pictures.

My aim is to encourage the beginner to this art with ideas for simple plaques, while for those who have had previous experience I hope to stimulate their imagination with some of my new and original ideas and designs for pictures, some of which are created with special occasions in mind.

Maureen Foster
Wilton, 1977

1 Plant materials and how to preserve them

The great single difference between an artist and a floral artist is the materials with which he works. The artist has to draw each shape and form, and then mix each colour to paint them. But for the floral artist, nature has these already prepared in the form of flowers, leaves, twigs and grasses, etc. It is necessary only to gather, preserve and store them, after which time they will be readily available whenever they are needed.

Plant material for pictures

What ever type of plant material we choose to use in our picture, I must emphasize the need to preserve a selection of different shapes and forms, and a variety of different sizes. So often I find a beginner with a small collection of preserved plant materials quite unable to create a design which pleases her because of the lack of variety.

Since the publication of my first book in 1973, I have continued to discover more and more plant materials that can be successfully preserved, many of them providing unusual and beautiful additions to preserved pictures. Together with many of my old favourites I have grouped these under the following headings: Seedheads and Cones, Grasses, Flowers, and Foliage. These lists by no means exhaust the limits of suitable plant materials: as you explore new ideas and new concepts for pictures I am convinced you will find more and more, just as I have.

Designing and making pictures in a variety of types, styles and sizes has made it possible for me to illustrate the use of a wide range of these plant materials. Readers requiring more detailed de-

scriptions of plant materials should see my book *Preserved Flowers – Practical Methods and Creative Uses*.

Seedheads and cones

Although lacking the bright colourings of flowers, these plant materials have much to commend them. In particular, they teach us to be aware of a special type of beauty – that of shape, form and texture. Seedheads and cones need little or no special preservation treatment. They are generally easy to find and collect, and easy to handle, making them ideal materials from which the beginner can create her first designs.

Although I have used picture frames with glass for some of the work illustrated in this section, glass is in no way essential. It does of course eliminate the necessity of making a recessed box, although even without glass a design will usually look more effective if it is recessed. The other advantage of glass is that it keeps the arrangement dust-free.

To list all the seedheads, both wild and cultivated, would be an almost endless task, but if you begin by collecting some of those shown you will soon learn to become aware of seedheads and seedpods. You can discover all kinds of interesting ones just by looking around your own or a friend's garden, when walking in the country or on holiday either in your own country or abroad (see Plate 5).

Spiky seedheads which consist of many smaller individual seedpods growing on the same stem can often be divided, and each tiny seedpod used separately for miniature pictures.

Seedheads are preserved by the air preserving method. See Fig. 5.

Above: PLATE 2 A FLOWERLESS DESIGN Here I have
concentrated on using contrasting shapes of plant materials.
The three flower-like shapes which form the centre of the
design are Hellebore orientalis; their petal-like forms are in
fact bracts. The insignificant flowers which bloomed in the
centre have faded and in their place tiny seed pods are just
beginning to form. The curious round clusters surrounding
the pods are individual clusters of bracts taken from a stem
of Ballota, and close by are clusters of Beech Nuts. The
dark disc shapes are leaves of the Eucalyptus Tree, their
darkness relieved by spiky seeds or grains of the Wild Oat
Grass. Other plant materials used are gleaming silvery discs
of Honesty and the crozier shapes of unfurled Fern fronds,
together with five different species of cultivated grasses,
Hare's Tail, Yellow Bristle Grass, Canary Grass, Quaking
Grass and a spike of Panicum violaceum. The background
is a coarse-textured linen-weave fabric.

PLATE 3 ADAM-STYLE FIRE SCREEN An Adam-style design
using mainly cones in a variety of shapes and sizes, with
just a few Poppy seedheads and gleaming silvery flower-like
bracts of the Wild Knapweed. Cones cut in half also make
unusual flower-like shapes and add interest to the design.
To make stems for the centre arrangement I used knobbly
branches of the Larch Tree. The frame is made from
natural wood which has been lightly stained; for the
background I used a piece of veneered plywood.

16

PLATE 4 A SELECTION OF SEEDHEADS 1. Wild Nipplewort
(Lapsana communis); 2. Wild Mugwort (Artemisia); 3.
Poppy (Papaver); 4. Wild Marjoram (Origanum vulgare); 5.
Lover-in-a-Mist (Nigella); 6. Grape Hyacinth (Muscari); 7.
Wild Mugwort (Artemisia); 8. Catananche; 9. Columbine
(Aquilegia); 10. Lavender; 11. Lily; 12. Evening Primrose
(Oenothera); 13. Wild Knapweed (Centaurea); 14.
Jerusalem Sage (Phlomis fruticosa). Other seedheads can be
seen in the pictures and plaques shown on pages 18–21.
Note: Clematis tangutica, the Wild Clematis, known also as
Old Man's Beard or Traveller's Joy (see Plate 21) produces
beautifully bearded seedheads. These should be gathered
before they turn fluffy and preserved by the glycerine
method, page 36.

PLATE 5 HOLIDAY MEMORIES FROM
ABROAD Preserving flowers while on
holiday abroad is not to be
recommended, mainly because of
the laws against picking many of the
protected wild flowers. There is also
the problem of taking a desiccant
with you, and of course once the
flowers are preserved, you have the
added problem of transporting them
home without damaging them – an
almost impossible task when coping
with the usual holiday baggage. Just
as much interest and excitement can
be had from searching and finding
unusual seedcases and pieces of
twisted bark, etc. (Dried plant
materials, including empty
seedcases, can be brought through
customs without any problems.) On
a recent holiday in the Canary
Islands I found at the base of a palm
tree some leaves which had been
restricted in their growth by a stone
wall. Their distorted shapes were
completely dried and parched by the
hot sun. Combined with Eucalyptus
and various other seedpods, and
mounted on a piece of veneered
plywood, this plaque will always
remind me of my holiday there.

PLATE 6 CONES A selection of cones, illustrating some of their many beautiful forms. Also shown in the top left-hand corner are Beech Nuts; they are ideal to use with cones but they must be preserved by the glycerine method while still tightly closed. After preserving, the husks will open but the nuts will remain intact.

Cones (or wooden flowers as I prefer to think of them) continue to fascinate me as much now as when I first began collecting them. They are ideal to use with seedheads for both plaques and pictures, particularly as both require the same type of background material, see page 40. They require nothing more than a brush-over to remove any clinging earth; and they can be collected and stored in boxes over an indefinite period, enabling you to build up a collection of different types, sizes and shadings which you will always have at hand for picture making. See Fig. 1 for wiring cones for use in swags, etc. Make sure the cones are quite dry before attempting to store them.

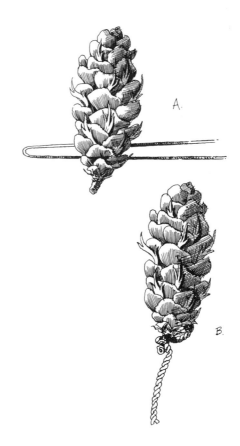

Fig. 1 Making wire stems for cones
(see page 20 for instructions)

Left : PLATE 7 DESIGN OF GRASSES A simple design using a selection of wild and cultivated grasses which have been intentionally bleached by standing them on a sunny windowsill. Their cream colouring is shown to advantage against a background of black cotton velvet. The cream patterned frame has just the right link with the grasses both in colour and texture. In addition to the grasses just a few florets of glycerined Moluccella add interest to the centre of the top and bottom design.

Opposite : PLATE 8 SEEDHEADS, CONES AND GRASSES ON A CARPET BACKGROUND Measuring 2 ft × 2 ft 6 in. (o·6 × o·75 m), this large old oak frame with its dull gilt slip is very much in keeping with the bold design of seedheads and cones. The somewhat unusual background consists of remnants of a bluish-grey carpet. Among the plant materials used were globular Leek seedheads, spiky Cardoon Thistleheads – their dead petals being pulled out when dry to reveal gleaming silvery flower-like centres, Poppy seedheads, long seed spikes of Wild Mullein, Barley, Wheat, Fir cones and glycerined Solomon's Seal and Mahonia leaves.

Making wire stems for cones (Fig. 1)

Use 19 gauge wires, which can be purchased in small packets from most florist's shops. For wiring tiny cones you may need to use a slightly finer gauge.

[A] Bend the wire in half and thread it between the bottom two rows of scales. Pull both ends of the wire tightly together until they become firmly embedded in the cone.

[B] Twist the ends of the wire round each other and then round the little stub of cone stem (this will help to support the cone). Continue twisting as shown.

Grasses

This section of plant materials includes both cultivated and wild grasses, together with the cereal grasses. Many are particularly dainty and graceful and provide excellent material for pictures almost without the addition of other plant materials; see Plate 7. When designing and making larger pictures, cereal grasses such as Barley and Wheat combine well with seedheads and cones; see Plate 8.

Wild grasses can be found in abundance everywhere, but they must be picked at the end of May or early in June. Grasses picked at this time will retain their beautiful green and mauve colourings, and will remain intact (which is of course essential for pictures). If gathered later, as the grasses turn golden and the seeds ripen and begin to fall, much of their beauty will be lost. A few grasses, for instance Marram Grass, produce their spiky inflorescence much later, but generally I find most of the more decorative species are among the early ones.

Many exceptionally decorative grasses can be grown at home in a small patch of garden – a packet of mixed seeds can produce many interesting forms. See the List of Suppliers at the end of the book.

Pick the grasses at the suggested time and preserve them using the air preserving method, page 32.

Flowers

When we select flowers to preserve for pictures, we will usually be concentrating on colour – colour blending of flowers within the picture, and the colouring of the picture as a whole in relationship to the room in which it is finally intended to hang. We must, however, not overlook the importance of contrasting shapes and forms. A picture composed entirely of solid round flowers or indeed a picture using only spiky flowers would be most uninteresting.

To provide a quick and easy reference and avoid confusion in preserving I have divided this group of plant materials into two sections: Section A – flowers which, due either to the characteristic formation or texture of their petals, can be preserved using the air preserving method, method 1, and Section B – flowers which need to be preserved in a desiccant, method 2.

SECTION A
Flowers, including flower-like bracts, that can be preserved in the same way as seedheads and grasses using the air preserving method, page 32 (see Plate 9)

(E) denotes everlasting flower

Colour	Flower	Chart No.
Blue-mauve	Delphinium	—
	Lavender	1
	Statice (E)	8
	Xeranthemum (E)	14
Red-purple	Heather (Ling)	9
	Helichrysum (E)	13
Pink	Acroclinium (Helipterum) (E)	2
	Helichrysum (E)	13
	Rhodanthe (E)	—
	Sedum	—
	Statice (E)	8
	Xeranthemum (E)	14
White–cream	Achillea (wild)	11
	Acroclinium (Helipterun) (E)	2
	Ammobium (E)	3
	Anaphalis (E) (Perennial plant)	—
	Moluccella (Bells of Ireland)	
	(Colour obtained after preserving may also be treated by the glycerine method, see page 36)	*see Plate 7*
	Gypsophila	—
	Helichrysum (E)	13
	Rhodanthe (E)	—
	Statice (E)	8
Yellow–gold	Achillea	4
	Solidago (Golden Rod)	6
	Helichrysum (E)	13
	Helipterum (E)	5
	Lonas (E)	—
	Statice (E)	8
Orange	Chinese Lantern (inflated calyx of the Cape Gooseberry)	*see Plate 36*
	Helichrysum (E)	13
Green	Alchemilla	12
	Ballota	7
	Woodsage (Wild)	10

Most flowers would just wither and die if you preserved them in the same way as seedheads and grasses, because of the size and texture of their petals. But there are a few flowers which will

PLATE 9 FLOWERS WHICH CAN BE PRESERVED BY THE AIR PRESERVING METHOD

respond reasonably successfully to this method of preserving and some are shown in Plate 9. They are flowers which consist of many tiny, closely packed florets. If, however, you want flowers of quality rather than quantity I would suggest treating them in the same way as the flowers described in Section B.

Plate 9 also shows a small group of flowers known as Everlastings or Immortelles. These require nothing more than the simple preservation treatment of method 1 because of the characteristic papery texture of their petals, which are not really petals at all, but bracts. These flowers are natives of Australia, Africa, Europe and the Mediterranean regions, and in Britain are cultivated from seed and treated only as half-hardy annuals. It is interesting to note that in 1731 Miller in his *Gardener's Dictionary* mentions two of these so-called everlasting flowers, Helichrysum, then referred to as Elichrysum or Eternal Flower, and Xeranthemums, of which Miller says, 'These were formerly much more cultivated in English Gardens than at present. Gardeners near London did cultivate in great plenty for their flowers which they brought to market in the winter season.'

This section also includes other plants which produce flower-like bracts and, although not classed as everlastings these will respond satisfactorily to the air preserving method.

SECTION B

A selection of flowers which can only be preserved in a desiccant using the method on page 32 (see Plate 10)

A number which is preceded by C refers to the flower as shown in the chart on page 25. A number preceded by P refers to the plate in which the flower can be seen used in either a plaque or picture.

Colour	Flower	Notes	Chart/Page No.
Blue	Anemone (St Brigid)		P19
	Cornflower		C6
	Delphinium		C18
	Forget-Me-Not (Myosotis)		C31
	Hydrangea	Individual flowers picked midsummer	—
	Larkspur		C28
	Love-in-a-Mist		C4
	Scabious		P30
	Hyacinth	Individual florets	—
Mauve-purple	Anemone (St Brigid)		—
	Cornflower		C6
	Dahlia	Decorative or pompon	C26
	Freesia		C12
	Lilac (Syringa)		C10
	Salpiglossis	Beautiful markings	—
	Salvia (Horminum)	Purple bracts	—
	Saxifrage (Mossy)	Pink changing to mauve	C16
	Hellebore (Orientalis)		C1
Red	Carnation		PXIV
	Cornflower		C6
	Dahlia	Decorative or pompon	PVIII
	Daisy (Bellis)	Cultivated form	C23
	Dianthus		—
	Fuchsia		P32
	Hawthorn		C27
	Hollyhock (Althaea rosea)	Single variety	PXIV
	Marigold	Deep bronze-red	C7
	Nicotiana		
	Peony (Tree Peony)		C2
	Roses	Miniature, Rambler and Florist Roses are particularly good (avoid fully open blooms)	C21
	Salpiglossis	Beautiful markings	—
Pink	Anemone (Japanese)		—
	Antirrhinum		
	Bergenia	Individual florets	C15
	Carnation		PIV
	Escallonia	Shrub	—
	Fuchsia		P32
	Hydrangea	Individual flowers picked midsummer	—
	Roses	Miniature, Rambler and Florist Roses are particularly good (avoid fully open blooms)	C21
	Salvia (Horminum)	Pink bracts	—
	Zinnia		C32
	Hyacinth	Individual florets	—
White-cream	Anemone (Japanese)		—
	Antirrhinum		—
	Carrot	Wild plant	
	Chamomile	Wild plant	—
	Cornflower		C6
	Daisy (Bellis)	Common lawn daisy	C23
	Deutzia	Shrub	C9

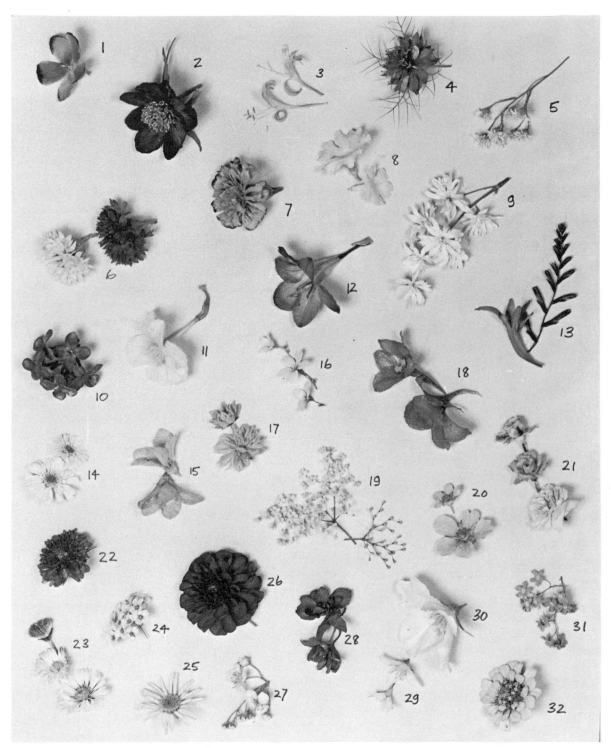

PLATE 10 FLOWERS WHICH CAN ONLY BE PRESERVED IN A
DESICCANT

Colour	Flower	Notes	Chart/Page No.
	Elder	Wild shrub	C19
	Feverfew		C14
	Hawthorn	Wild shrub	C27
	Hollyhock (Althaea rosea)	Single variety	P31
	Honeysuckle (Lonicera)		C3
	Lily of the Valley		PV
	Philadelphus	Single variety	C29–30
	Saxifrage (Mossy)		C16
	Scabious		P30
	Spiraea	Shrub	C24
	Zinnia		C32
	Passion Flower		—
Yellow-gold	Anthemis		C25
	Antirrhinum		PIX
	Carnation		—
	Cytisus		PIX
	Dahlia	Decorative or pompon	C26
	Golden Rod (Solidago)		C9
	Kerria japonica	Shrub	C17
	Marigold		C7
	Narcissus		C11
	Peony (Tree Peony)		C2
	Potentilla	Shrub	C20
	Roses	Florist Roses are particularly good	—
	Salpiglossis		—
	Solidaster		C5
	Zinnia		PXV
Orange	Carnation		—
	Dahlia		PIX
	Day Lily		PIX
	Marigold		PIX
	Montbretia	Buds are particularly useful	PXV
	Zinnia		—
Green	Hellebore foetidus		C1
	Hydrangea	Individual florets	—
	Nicotiana		—

This is not just a list of flowers but, to be more accurate, flowers which I consider suitable and which will preserve to a high enough standard for pictures. To a certain degree it is possible to preserve all flowers, but many become too limp after preserving, losing much of their characteristic form, while others with very thick and fleshy petals cannot be preserved quickly enough to retain their true colourings.

Much of the success of a perfect preserved flower is achieved by starting with a perfect fresh flower. Check flowers and discard for the following reasons:

1. Insect holes.
2. Distortion of petals.
3. Unsightly marks, often caused by rain.
4. Loss of colour from sun bleaching.

5. Most important – flowers that are past their peak of perfection. At this stage in their growth the seeds are beginning to form and it is most likely that, if preserved, the petals will fall out.

Never preserve flowers that are retaining moisture. After rain or early morning dew allow time for flowers to dry out completely, remembering that some flowers with particularly dense centres may take a considerable time. It is most important that flowers intended for preservation should hold no trace of moisture, as this will cause the petals to turn brown.

Most flowers intended for use in plaques and pictures only need to retain a very short length of stem, but flowers which are being preserved for the type of plaques and swags shown in Plates 18 and 19 will naturally require a longer stem. Many flowers such as Delphiniums and Rambler Roses have stems which retain their firmness sufficiently well after preserving for this type of work, but even with these it may be necessary to lengthen the stem; see Figs 2B and C. Flowers with very fleshy stems such as Anemones or Dahlias will need to be

Fig. 2 Wiring flowers

wired. This must be carried out before preserving while the centre of the flower is soft; if left until after preserving it will be impossible to push a wire through without the flower disintegrating; see Fig. 2A for wiring.

Wiring flowers (Fig. 2)

Use 30 gauge wires for small flowers and 20 gauge for larger flowers (obtainable from most florist's shops).

Take a wire and pierce the centre of the flower with it, bringing it out where the natural stem begins. Discard the flower's natural stem and pull the wire through just far enough for the tip to become embedded in the centre of the flower. During the preserving process there will be a slight corrosion between the wire and the flower ensuring that when your flower is preserved the stem will remain firm and secure.

Lengthening a stem (Figs 2B and C)

Use wires as described above. Place a wire to overlap the flower's natural stem by approximately $1\frac{1}{2}$ in. (3.5 cm) and bind the two together with either florist's silver reel wire or fuse wire.

Alternatively, collect and dry strong hollow plant stalks of various sizes and thicknesses. Select a suitable stalk and push the stem of your flower into it. Squeeze a blob of glue on to the tip of the stem first to ensure firm fixing.

Binding false stems with plastic florist's tape gives a professional finish but is time-consuming. I find that in plaques and swags, as opposed to flower arrangements, stems are in any case usually covered by other flowers or foliage.

Readers of my first book will already be aware of my definition of flower types, which I think in many ways may be the result of my early training as a corsetiere. Knowing as I do that people fall into one of a number of figure types regarding shape and form, I found myself looking at flowers in the same way to create simple methods of positioning during the preserving process.

Flower types – a guide to positioning flowers for preserving (Fig. 3)

[A] *Flat or rounded forms.* These include flowers with flat faces such as lawn Daises, and rounded globular flowers such as decorative Dahlias.

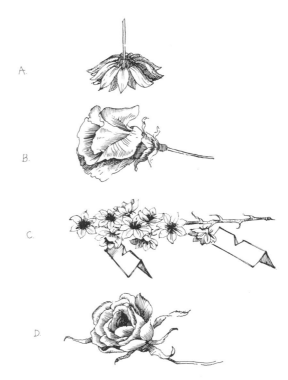

Fig. 3 Flower types – a guide to positioning flowers for preserving

[B] *Trumpet or bell forms.* These include flowers such as the single Hollyhock. Fill the trumpet shape with desiccant first, before covering. This is also an ideal position for partially open flat flower forms, also for those with pronounced stamens.

[C] *Spiky forms.* Usually flowers which are made up of many smaller florets, all growing round a single spiky stem; the Delphinium is a good example. Use two supports of cardboard as shown to avoid the underneath florets becoming squashed.

[D] *Pointed forms.* Double flowers, usually with many large, closely packed overlapping petals such as Roses, allow the desiccant to fall in between the petals. This is also a useful alternative positioning for flat or rounded forms.

Preserving special occasion flowers

Most of us are sentimental at heart, so why not make a picture as a keepsake of a special occasion such as a friend's wedding. I always wish I could

have kept the flowers from my wedding bouquet, but unfortunately at that time flower preserving had not entered my life. Even when I did start to preserve flowers it wasn't until one day when someone telephoned to say, 'Can you tell me how to preserve flowers – my daughter is getting married tomorrow and I would like to be able to preserve the flowers from her bouquet, as a keepsake', that this idea occurred to me. See Colour Plates IV and V.

Keep the bouquet in a cool place after the event until it can be dealt with for preserving. *Don't* spray it with water or float it in water, as waterlogged flowers will turn brown in the centre if you try to preserve them while they are still wet. Flowers such as Carnations and Roses hold water deep down between their petals for a long time, and it is almost impossible to dry them out again completely before they start to wither.

If the flowers are wired the stem of the bouquet will consist of a collection of wires instead of flower stems. Standing this kind of bouquet in water will have no effect on the lasting qualities, and may even be harmful if the flowers themselves come into contact with the water.

In a relatively new approach to wedding bouquets the flowers are arranged naturally (without wiring) in a container holding wet Oasis (plastic foam). Not only does this greatly extend the life of the flowers but it eliminates the rather tedious job of unwiring each flower before it is preserved.

Foliage

Whenever we use flowers we need foliage, even in preserved flower pictures, for foliage colourings act as a natural foil for the flowers, and help to give them visual unity. The wide variety of shapes and forms will add interest to a picture.

Opposite: PLATE 11 AUTUMN LEAVES The interesting and unusual corkscrew-like branch of the contorted Hazel combined with a few preserved autumn leaves makes a simple yet interesting picture. The branch is in fact made up from several individual pieces of branch which I carefully arranged and stuck in position to resemble one complete branch. The background of natural jute wall-covering adds textural interest to the picture. The narrow wooden frame is painted in a shade of brown which picks up the earthy colouring of the branch, and an edging of off-white paint prevents the frame appearing too dark and drab.

Many small, compact perennial plants produce exceptionally attractive foliage, enabling people with only the smallest of gardens to grow them. If I only had enough space to grow either flowers or foliage plants, I would always choose foliage. We can buy flowers, but it is very difficult to find a selection of foliage at a florist. It is also possible to find many interesting leaves during a walk in the countryside.

With the choice of three different methods it is possible to preserve foliage for pictures at any time of the year and at any stage of growth, but it is important to understand the effect each method has on the different foliages, often making one method more suitable than another.

The desiccant method of preserving is one which I use only for small leaves. Almost any leaf can be preserved in this way, but the brittleness after preserving makes large leaves too easily broken. Follow the basic directions given for flower preserving on page 34. Take care as you put the leaves into the desiccant to retain the shape and form of each one.

This is the ideal way of preserving the beautiful colouring of autumn leaves, in fact (apart from pressing) it is the only way.

A selection of small leaves suitable for preserving by the desiccant method (see **Plate 12**)

Colour	Foliage	Notes	Chart No.
Green	Alchemilla mollis	Young leaves	16
	Aquilegia	Large leaves may be divided into individual segments	—
	Cupressus		—
	Cyclamen neapolitanum	Beautifully marbled leaves	—
	Epimedium	Small young leaves	24
	Ferns (including unfurled fronds)	If ferns are large use individual segments	14–15
	Kerria japonica		6
	Rose Leaves	From Miniature or Rambler Roses	20
	Tellima	Young leaves have beautiful veining	19
	Tree Peony	Small leaves	2

Note: All green leaves tend to turn to a yellow-green after a few years but still remain attractive.

Colour	Foliage	Notes	Chart No.	Colour	Foliage	Notes	Chart No.
Silver or grey-green	Acaena		3		Sage		13
					Senecio (greyi)		1
	Alchemilla alpina	Reverse, using undersides	—	Red-plum	Clematis montana		4
	Ballota		5		Herb Robert	Wild	23
	Chrysanthemum Haradjanii		9		Rose Leaves		20
	Cineraria maritma (Senecio cineraria)		21		Also numerous autumn leaves		
	Cupressus		—	Yellow	Aralia		18
	Dianthus	Ideal when small spikes are needed	11		Cupressus		10
					Elaeagnus	Small leaves (variegated)	25
	Potentilla anserina	Wild	—		Parsley		17
	Rue	Leaves may be divided into individual segments	8		Rose Leaves		—
					Also numerous autumn leaves		

Above: Plate 13 A Selection of Foliage Suitable for Preserving by the Glycerine Method, Chosen to Illustrate a Variety of Shapes 1. Wild Hawthorn or May (Crataegus monogyna); 2. Rosemary (Rosmarinus); 3. Elaeagnus pungens; 4. Forsythia; 5. Cupressus; 6. Cotoneaster horizontalis; 7. Cineraria maritma (Senecio cineraria); 8. Rose; 9. Beech (Fagus sylvatica); 10. Escallonia; 11. Oak (Quercus).

Opposite: Plate 12 A selection of Small Leaves suitable for Preserving by the Desiccant Method

The glycerine method (see page 36) will be found to be particularly suitable for foliage which is intended for use in plaques, as the glycerine mixture is absorbed into the leaves, keeping them soft and pliable, but unlike the other methods there will be a change in colour. With some you may notice only a slight change from perhaps mid-green to a darker green, but others can change from green to a really dark brown or even a light sherry colour.

I welcome these changes, for they help to extend the range of colourings in foliage, making it possible to create colour schemes for pictures and plaques that cannot be achieved in arrangements of fresh flowers.

Other foliage preserved by this method and used in my plaques and pictures but not shown in this chart includes Camellia, Choisya, Eucalyptus, Laurel (Laurus nobilis), Magnolia, Mahonia, Pittosporum tenuifolium, Solomon's Seal

Fig. 4 Wiring individual sturdy leaves for plaques

As most evergreen foliage comes from trees and shrubs you can preserve complete sprays but not more than 1½ to 2 ft (0.5 m) in length. For use in plaques and swags these can be divided after preserving, and if necessary some leaves can be wired individually; see Fig. 4.

Wiring individual sturdy leaves for plaques (Fig. 4)
Using a heavy gauge florist's wire pierce the base of the leaf just above the existing stub of stem. Pull the end of wire down, twisting it round the leaf stem and the other end of wire at the same time, as shown. Bind with plastic floral tape if the wire is likely to be seen, but usually it becomes hidden by other plant materials.

(Polygonatum odoratum), and Garrya elliptica. other plant materials suitable for preserving by this method include sprays of Beech Nuts, Moluccella (bracts) and Clematis seedheads, both wild and cultivated. Most seedheads will in fact respond well to this method of preserving as an alternative to natural preserving. The advantage is of course that they will be more pliable and therefore easier to handle. *Note:* Seedheads preserved by this method *will* change colour.

Most deciduous foliage suitable for preserving comes from trees and shrubs, but some of the firmer-textured herbaceous leaves can also be preserved in this way provided they have stems which are firm enough not to collapse while standing in the glycerine mixture. If, however, you do experience problems, use the alternative glycerine method B for this type of herbaceous leaf.

To obtain successful results with deciduous foliage gather it between July and late September. Before July the leaves will be too young and soft, and towards the end of September the flow of sap to the leaves will decline, so too will the necessary speed absorption of the glycerine mixture.

Evergreen foliage is so useful for swags such as the one shown in Plate 21, as the leaves are particularly firm and tough, often with quite a leathery texture. We have the advantage of being able to gather evergreen foliage during the winter months when deciduous foliage is unobtainable, but do avoid picking the new leaves which form in the spring.

Methods of preserving plant material

Method 1 – Air preserving (Fig. 5)
Choose a convenient place which is cool, dry, airy and dark, and preserve in whichever of the following two ways is suitable to the type of plant material.

[A] The ideal way of preserving seedheads and certain flowers (see the list on page 22). Tie them and hang in small bunches. The plastic-covered wires used for deep-freeze packs are ideal for this purpose as the bunches can easily be tightened if necessary as the stems dry.

[B] An alternative way of preserving by the air method providing the plant materials have firm stems, but in particular it is the ideal way to preserve grasses, encouraging them to develop interesting curves.

Method 2 – Preserving with desiccants
A desiccant is a substance which absorbs moisture. Sand, alum, cornmeal, starch, detergent powders, borax powder and silica gel are among the products used for this purpose. It is my opinion, however, that the importance of a desiccant is not only to absorb the moisture but to absorb it quickly, enabling the natural colouring of the flower to be retained. A slow preserving process usually results in loss of colour. Basically the

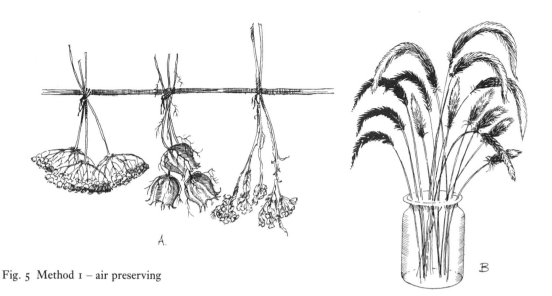

Fig. 5 Method 1 – air preserving

process of preserving a flower is to cover it completely with the desiccant; therefore the degree of success achieved in preserving the form and contours of various flowers partly relies on the weight and density of the desiccant you choose.

The two products which I have used with the greatest success are borax and silica gel, but as with all art and craft products it is advisable to experiment for yourself. In this way you will find a desiccant that is right for you, particularly as the conditions available for preserving may also have to be taken into consideration. As a guide for readers who have no previous experience with desiccants I will explain the advantages and disadvantages I have found with borax and silica gel.

Borax
Household borax is inexpensive and easily obtainable from chemists and drugstores. It can be used over and over again indefinitely. Being in powder form, it is light in weight and therefore particularly suitable for small flowers. Its disadvantages are (a) it needs a dry heat (preferably an airing cupboard) to obtain successful results; and (b) it has a tendency to cling to some flowers, which is particularly noticeable on dark-coloured flowers. A small artist's paintbrush will quickly remove any clinging borax.

Silica gel
Silica gel is used industrially for absorbing moisture. It looks rather like granulated sugar. Although more expensive than borax, silica gel has the advantage that heat is unnecessary as it absorbs up to fifty per cent of its own weight of moisture without the crystals appearing or feeling wet. They must, however, be dried out when they have absorbed their maximum amount of moisture. Treated in this way, these too can be used indefinitely.

It is often difficult to obtain a suitable grade of crystals as the standard grade ones sold by chemists and drugstores are usually too large even for sturdy flowers. The intense hardness of these crystals makes them unsuitable for crushing, even in an electric grinder. It is best to buy crystals sold especially for flower preserving. These are sold under a variety of trade names and usually contain silica gel but of a more refined grade than the standard crystals. These special crystals contain a quantity of blue crystals which turn pink when the maximum amount of water has been absorbed. When they reach this stage it becomes necessary to reactivate them (see end of method for details).

On account of the difficulty I have experienced in obtaining and recommending retail sources for flower preserving crystals I have recently arranged for my own flower preserving crystals to be made

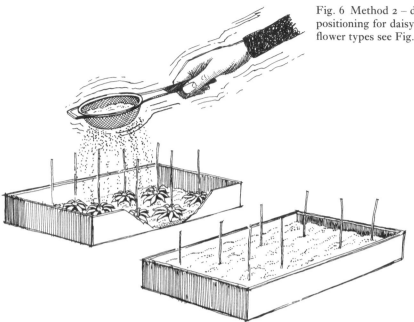

Fig. 6 Method 2 – desiccant preserving with borax: basic positioning for daisy-type flowers. For positioning of other flower types see Fig. 3 (page 27)

available. See the List of Suppliers at the end of the book.

Desiccant method (Figs 6 and 7)

I will describe this method for borax only – the method for silica gel is the same, with the few differences listed on page 35. First make sure the borax is completely dry. It should run through your fingers like table salt. If it clings together and is inclined to be lumpy spread it out on a shallow tray and leave it in a warm place to dry, such as an airing cupboard. Unless the borax is quite dry your flowers will not preserve successfully.

1. Find a suitable-sized cardboard box; this will depend entirely on the flowers you are preserving. Shoe boxes are ideal for large flowers, while writing paper boxes, chocolate boxes, or in fact any box with a depth of 2 in. (5 cm) or so will be suitable for tiny flowers such as Daisies or Forget-me-nots.

1. Spoon borax into an empty box to a depth of about 1 in. (2.5 cm).

3. For positioning and covering your flowers see Figs 6 and 7. Preserve flowers of one type only in each box. This important step in the preserving process determines the shape and form of your flowers after preserving. Petals crumpled or displaced at this stage will remain that way.

4. Place your box of flowers without a lid in a warm airing cupboard.

5. The amount of time taken for flowers to preserve varies according to type, depending mainly on the thickness and density of their petals. As a guide, a small Lawn Daisy should be ready in about twenty-four hours, while a Rose or Dahlia will take four to five days.

6. To test if the flowers are ready, lift out *one flower only*, hold it to your ear and give it a gentle flick; if it sounds papery it is ready. This will avoid wasting a whole batch of flowers that may not be completely preserved, as partially preserved flowers cannot easily be returned to the borax.

7. Flowers that have been preserved in this way require extra-special care during handling and storage. When first removed they are extremely fragile and should be left at room temperature for several hours before cleaning and storing, by which time they will have absorbed just enough of the natural moisture from the atmosphere to render them more manageable. Have ready some small pots containing chunks of dry Oasis or sand in which to stand flowers with natural or false stems, and a tray on which to lay stemless flowers.

8. To clean the flowers, gently flick the stem, or in

the case of stemless flowers, the back of each flower, to remove excess desiccant. Now examine each flower and if necessary take a small, soft artist's paintbrush and carefully brush each flower free from any remaining particles.

9. The ideal way to store flowers with natural or false stems is to leave them in the pots of dry Oasis. Stand these pots in a suitable-sized box with an airtight lid. Keep a small, open container of silica gel or preserving crystals in the box. These will absorb any moisture and, if changed regularly, will keep your flowers free from damp or humidity.

I use an office filing cabinet for storing stemless flowers. It is a good idea to line the drawers to prevent the flowers shaking about as each drawer is removed during your work. Shallow boxes with lids can also be used. Lay the flowers down side by side – *don't* pile them one on top of the other, as this can be disastrous for any preserved flower. Keep small containers of silica gel crystals in each drawer in the same way as described for flowers with stems.

I find it most successful to keep flowers of each colour separately, i.e. all the blue shades together in one drawer, all the yellows in another, and so on. This is for two reasons: (a) it is so much easier when working out colour schemes, and (b) it is easy to see at a glance which colours are running low.

For silica gel, follow the instructions for borax preserving with the following amendments.

1. Use tins or polythene containers with airtight lids. Coffee tins, biscuit tins and polythene ice cream containers are all suitable, depending on the size and quantity of flowers you intend to preserve in each batch.

2. Silica gel and flower preserving crystals are too coarse to pass through a sieve. They must be carefully sprinkled around your flowers using a spoon (see Fig. 7). As you continue to sprinkle them around each flower they will gradually fall in between the petals until finally all the flowers are

Fig. 7 Method 2 – desiccant preserving with silica gel. For positioning of other flower types see Fig. 3 (page 27)

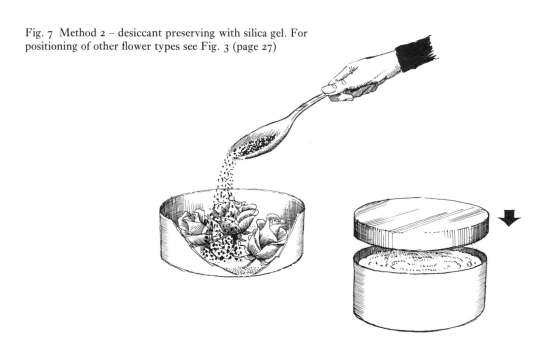

completely covered. Do not let the crystals fall heavily on top of your flowers as this will make them lose shape and form.

Reactivating preserving crystals

This simple process is necessary from time to time as the crystals become fully charged with their maximum amount of moisture. The method is as follows. Pour the crystals into a shallow pan and leave in a warm oven set at 250°F for fifteen to thirty minutes, or until the blue colour reappears, indicating that the crystals are dry. An occasional stir helps to ensure even drying. Some crystals on the market show moisture content in other ways, therefore always check the instructions given with the crystals you are using.

Colour retention

When using the desiccant method of preserving flowers I find that I usually achieve ninety per cent colour retention success, providing of course that the flowers are kept under correct conditions both during storage and in the finished picture; see pages 35 and 78. However I feel it is worth mentioning the types and colours of flowers with which I have experienced problems.

With the exception of certain species of the Dendrobium Orchid I find other members of the Orchid family lose most of their colour during the preserving process, resulting in shades of brown to fawn. This is caused by the length of time they take to preserve (about three weeks) owing to the high water content of their thick, fleshy petals. They may, however, be considered worth preserving for sentimental reasons as they do retain their shape and form beautifully. Other flowers with very fleshy petals such as Magnolias and Begonias present similar problems.

I have always found that yellow flowers such as Marigolds and Tree Peonies keep their colour beautifully, while others such as Winter Jasmine and Buttercups fade to almost pure white in a very short time. The texture of Marigold and Tree Peony petals is very different from that of the Jasmine and Buttercup, which may well be the reason.

Red flowers which have a yellow undertone keep their colour well, but flowers with a blue undertone darken quite considerably. This of course does not mean that they are not worth preserving, but you must be prepared to accept, for example, that a lovely ruby red Rose will turn to a very dark shade of red, sometimes almost black. Mauve and purple shades react in a similar way.

True blue flowers such as Delphiniums or Cornflowers retain their true blue colouring exceptionally well.

Many white flowers tend to turn cream after they are preserved. This never worries me unduly, in fact I sometimes feel they blend in to colour schemes even better than pure white flowers. Among these are Lily of the Valley and the shrubby Spirea.

Method 3A and B – Glycerine method
(Fig. 8)

Method A

1. Add one part of glycerine to two parts of hot water, and mix well together until the mixture looks clear. Glycerine is heavier than water, and unless you mix thoroughly the glycerine will remain at the bottom and only the water will be taken up by the foliage.

2. Pour this mixture into a container which is just large enough to take your foliage to a depth of about 3 in. (8 cm). For support, stand this container inside a larger container such as a bucket which is heavy enough to prevent the foliage toppling over. The stems of woody foliage should be split and put into the hot mixture directly they are picked. If left to stand in plain water first the foliage will become saturated and may not be able to take up the glycerine mixture before the top leaves begin to curl.

3. Leave standing in a cool, dry, dark place. The preserving time will vary with the different foliage, depending of course on the thickness and texture of the leaves. As a general guide foliage with small, thin-textured leaves, such as Cotoneaster and Escallonia, will be ready within a few days, while foliage with tough leathery leaves such as Elaeagnus or Laurel will take several weeks. It is interesting to watch the glycerine mixture travelling through

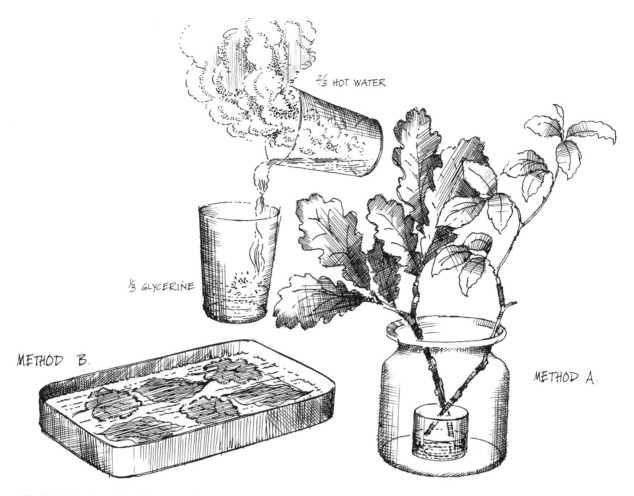

² ⁄₃ HOT WATER

⅓ GLYCERINE

METHOD B.

METHOD A.

Fig. 8 Method 3 – glycerine preserving

the leaves and see the gradual change of colour. Keep a spray of the same foliage in a jar of water to compare them. This way you will know immediately the foliage is completely preserved.

4. Over-glycerining foliage is a common fault, resulting in excess mixture seeping out through the leaves. If this occurs wipe away the sticky moisture with a soft cloth and then swish the foliage in warm water. Immediately pat dry with a clean, dry cloth and keep in a warm place until completely dry before storing.

Method B (Immersion)
For individual leaves with soft stems.

Lay your leaves in a suitable size dish in a non-rusting material such as glass or plastic. Prepare a glycerine mixture as described for Method A, and pour this over the leaves making sure they are completely covered. It may be necessary to weight them lightly to keep them submerged. Lift the leaves from time to time to see when the mixture has been absorbed into the whole leaf. Clean, wash and dry leaves as described in point 4 above.

2 Preparation for a plaque or picture

Creating the right picture for the right setting

A plaque or picture can of course be created purely for the fun and enjoyment of blending a collection of plant materials to form a pleasing design, but a much greater sense of interest and fulfilment can be achieved by planning and designing a picture for a particular purpose. This could be a present for a special occasion, or, if the picture is for your own home or a friend's home, select a room and position in which you want the picture to hang and then concentrate on designing something that is not only pleasing to you or your friend, but that will harmonize with the decor of the room. To do this will involve careful consideration of the following points: (a) type and colouring of plant materials; (b) type and colouring of background material; (c) type and colouring of frame (if used).

In the following pages I have dealt individually with each aspect of creating a picture. It is not always necessary to keep to this order of working, but the beginner will find it helpful to do so.

Inspiration

As you progress you will probably find, as I do, that the inspiration to create a picture can come from a wide range of things. It may be a piece of material with an interesting texture or an unusual colour, a few special flowers that are a keepsake of an important occasion, or maybe an interesting old frame found in a junk shop.

The important point to remember is that, whichever you start with – plant materials, background material or frame – you must aim to create a harmonious effect, and a sense of unity in your finished picture, so that each part appears to belong to the picture as a whole.

Planning a colour scheme

Although we may not always be aware of it, colour affects us all in many ways: not only in pictures, but in every walk of life – the furnishings in our homes, the clothes we wear, even the foods we eat often have artifical colouring added to make them more appealing. People often react to colour more than to design. Therefore a bad design may often pass unnoticed if it has an outstanding blend of colour. The first comment I so often receive when someone is looking at a picture is, 'I like the colouring', or, when looking at several different pictures, 'I prefer that colouring'. Colour is a very personal thing. We nearly all have a favourite colour which tends to reflect in our choice of clothes and furnishings and even our car.

Colour in plant materials
Colour blending is important – not only in flower pictures, but in pictures created from seedheads, cones and twigs, because although in such pictures the plant materials are basically all brown, the variation of tints and tones is tremendous, ranging from a pale cream through to a dark brown which quite often is almost black. Look at a collection of fir cones and see the enormous range of tints and tones of brown. Beginners tend to preserve a

PLATE 14 A BEGINNER'S FIRST PICTURE Designed and made
as a first picture during an evening class session. You will
see that light-coloured flowers have been successfully
combined with a pale-coloured fabric by introducing dark
foliage.

selection of flowers, only to find when they begin to make a picture that they have not preserved enough tints and shades of one colour, or enough colours which blend harmoniously together. This of course will result in a very unco-ordinated effect in the finished picture.

How then can we learn to achieve a pleasing and effective blend of colour? One way would be to group together little collections of preserved flowers in combinations of different colours, and then gradually eliminate colours which appear wrong to your eye and replace these with other colours, until you have a collection of colours which please you. Another way is to look at floral upholstery materials, particularly in the more expensive ranges, where you will find especially lovely colour schemes which you can recreate with preserved flowers. While doing this, keep in mind the colouring of your room and make quite sure you choose a colour scheme that will harmonize with it.

Choosing a background

Choosing a background for your plant materials is a very important step in making a plaque or picture, because of course there will be such a close association between the two. In fact, unless you choose a suitable background, you will find it very difficult to finish up with an attractive picture. I cannot emphasize too strongly how a wrongly chosen background can completely ruin the effect of a picture, just by being too light or too dark, the wrong colour, or even the wrong type of material. But the right background will compliment and enhance your finished picture.

Types of background materials

When using a particular type of plant material some background materials of course are more suitable than others. Understanding the correct use of different background materials really comes with experience in handling them and observing the effect they have on various plant materials. However, to guide the beginner, the following notes combined with the illustrations of pictures will, I hope, help to explain a little of the importance of background material in the presentation and effect of a finished picture. As a

general guide wood, hessian, neutral colour felts and linen-textured wall coverings are particularly in keeping with designs of cones and seedheads, while furnishing velvet, fine linen, satin, rayon and the jewel-coloured felts are more suited to designs of delicate desiccant-preserved flowers and foliage.

Some background materials will be more in keeping than others with your room, which of course will govern to a certain extent the type of plant materials you use. For example, a coarse hessian or wood is very much more suited to a modern room setting than a traditional one, where a fine linen, silk or velvet might be more appropriate. Whatever type of material you choose, avoid using patterned materials, with the exception of lace, as these will detact from the plant materials and result in a confusing picture.

Colour

Choosing the colour of your background will usually be restricted by your furnishing colours. While it is not necessary to match the colourings already used in your room, it is of course important to choose a colour which will harmonize. As a good general guide I use light colour backgrounds for dark plant materials, and dark colour backgrounds for light plant materials. But this does not mean, for instance, that you cannot use dark flowers in a design which is set on a dark background providing the dark flowers are surrounded by flowers of a lighter colour, or *vice versa*.

Preserved flowers have a rather matt appearance, due to the lack of moisture content which gives fresh flowers their sheen. For this reason I usually find it safer to avoid a background of brightly coloured material. The one exception occurs when you are working with a design of white flowers. White is a very strong colour, and a design composed almost entirely of white flowers is often shown to advantage against a bright background, particularly bright blue or bright green, providing the area of background colour which surrounds the design is not too great.

Try little groupings of your plant materials on different colour backgrounds, or observe the background colourings used in the furnishing fabric you were using as a guide for design colourings. Note the different effects obtained by placing the same colour groupings of plant materials on different colour backgrounds, and see how each background appears to change the colours of your flowers.

Texture

Choose your background not only for its colour value, but for its texture, too. For the pictures illustrated in this book I used many different textured materials as backgrounds. Let's look at a few of them.

Plate 8. Normally with smaller pictures using this type of frame and plant materials I would choose hessian or a similar type of fabric as a background, but with such a large frame I felt I could use something with a stronger texture. While I was crouched in my usual position on the carpet surrounded by seedheads, etc, trying to decide just what to use, I suddenly decided this was it – a carpet background. My own carpet – a bluish-grey twisted pile – seemed ideal, and after a hunt in the loft I was able to find enough pieces.

In contrast, see Plate 31. When preserved, the texture of these Hollyhocks is very much like tissue paper, and therefore unlike the large seedheads and cones they require a background of a more delicate fabric. After experimenting with several different types I finally used a piece of old lime green satin which not only picked up the lovely lime green colouring of the Hollyhock centres, but was also just the right texture, having a slight sheen which emphasized the opaque white of the Hollyhocks.

Experimentation

The beginner will learn much from experimenting with plant materials and backgrounds. Gather together small collections of different types of plant materials, and try them against backgrounds of different colours and textures. Even if you have to go round the shops or market to do this, it will be well worth while. You will observe that against one background your plant materials will perhaps look dull and lost, while against another they will be thrown into relief and look alive.

3 Making plaques and swags

For ease of reference my definition of plaques and swags is that these are without recessed frames. Plaques are ideal for the beginner to make, but it is important to realize that without a glazed recessed frame to protect the plant materials, preserved flowers in particular will have a limited life. Fig. 9 shows how to make an easy but attractive plaque which is illustrated in Plate 15. Plate 16 shows another plaque made in a similar manner. Details of a free-standing plaque are given in Fig. 10.

Cake doyley plaque (Fig. 9, Plate 15)

[A] With a pencil, lightly draw an oval shape on a piece of coloured card. To make an attractive edging to your plaque arrange and stick cut-out motifs from a cake doyley around this line.

[B] Leave until the paste or glue is dry, and then using a pair of pointed scissors cut round the outside edge of the doyley motifs.

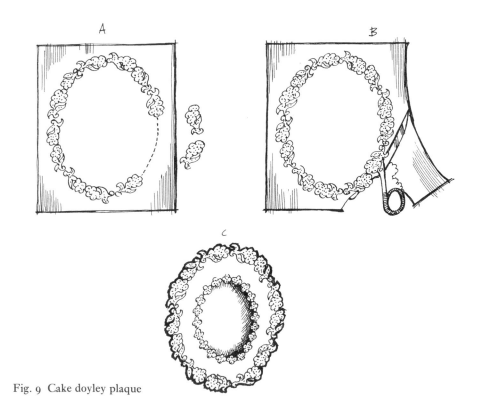

Fig. 9 Cake doyley plaque

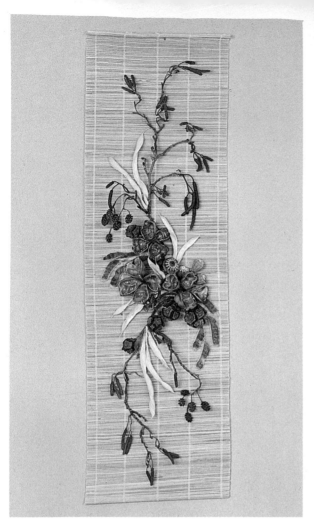

PLATE II ROLLING PIN 2

PLATE III VELVET AND LACE FRAMED PICTURE Slightly larger than the picture shown in Plate 25, this frame is covered with velvet and lined with cream lace. The flowers, in shades of mauve and purple, include miniature Zinnias, Forget-Me-Nots, Bergenia and Lilac, with contrasting gold Zinnias and Marigold buds.

PLATE I CONES AND SEEDPODS Texture, shape, form and colour are all essential to the success of this type of design. Branches from the Wild Alder Tree form the outline, with their tiny catkins and cones which are found together in December to January. The three flower shapes in the centre are seedcases of the Tree Peony, and surrounding them are cones of the Cupressus Tree and seedpods of the Oriental Poppy, with a few velvety seedpods of the herbaceous Lupin. Needing something light in colour and spiky, to contrast with the more solid forms, I turned to the kitchen garden. Here I found lovely twisted seedcases of the Dwarf French Bean. After the long hot summer of 1975 they were bleached to a beautiful shade of cream, just what I was looking for, which proves that good hunting grounds for plant materials are everywhere, and often in the most unlikely places.

Opposite: PLATE 16 GRANDMA'S BIRTHDAY PLAQUE Made in a similar way to the cake doyley plaque shown in Plate 15 (see Fig. 9), but instead of fixing lace to the plaque edge as described for the main doyley motifs, gather and fix the lace to the edge of the central padded shape as suggested for the smaller doyley motifs.

PLATE IV A KEEPSAKE Preserved for sentimental reasons, a selection from a bouquet of flowers given by a young man to his girlfriend. Pink and blue Larkspur, shades of pink Carnations, florets of yellow Gladiolus and a single red Rose.

PLATE V FLOWERS FROM A BRIDE'S BOUQUET A lasting visual memory of that special day! Roses, Freesia and Lily of the Valley arranged with their own foliage.

PLATE VI VALENTINE PICTURE A design in which
I used a selection of Everlasting Flowers in
shades of purple, mauve and cerise to
complement the rich purple velvet covering of the
box frame. As a background for the flowers I
lined my box with white satin (see Fig. 15 for
details of making and covering the box shape).

PLATE VII OVAL PICTURE IN MOULDED PLASTER
FRAME This oval picture is one of a pair commissioned by a famous builder of model ships. Both
pictures have backgrounds of deep blue velvet;
only the flowers differ. This one is composed of
shades of blue and cream, while the other one
(not illustrated) has pink and cream flowers.

PLATE VIII ANTIQUE-FRAMED PICTURE A beautiful antique gilt frame, somewhat heavier than the one used in Plate 14, with plant materials and background fabric chosen accordingly. To complement my flowers I chose a rich brown velvet. The mixed colourings of Dahlias, Helichrysums, Tree Peonies, Achillea, Anthemis, Hollyhocks, Spiraea and Delphiniums, with foliage in shades of cream and gold, blend together to create a harmonious effect with the frame.

PLATE IX STUDY IN GOLD AND ORANGE This beautiful antique frame is one of a pair I found in a junk shop: they had several pieces of moulding missing and the paintwork looked rather shabby. However I quickly set to work and with a little time and patience the frames were soon restored and ready for use, see page 71. A remnant of lime green dupion was the exact colour I needed for a background as it blended so beautifully with the greenish colourings of the frame. My choice of flowers in shades of yellow, gold and orange included Dahlias, Helichrysum, Freesias, Marigolds, Crystus, Antirrhinums and a Lily with brown and green foliage, together with a few ferns sprayed with gold paint (this is something I seldom do but somehow they just seemed right, and of course they picked up the gilt moulding of the frame).

PLATES X, XI, XII ADAM-STYLE PICTURES Based on the
designs of Robert Adam, these are three pictures which
illustrate how to use the many tiny flowers, florets and
leaves which can be preserved by the desiccant method.
The colours of the flowers and leaves in each picture were
chosen to tone with the colour of the velvet background. In
PLATE X I used shades of pink, mauve and cream with silver
leaves on deep plum coloured velvet. Flowers include
Miniature Roses, Acrocinium, Hawthorn, Daisies, shrubby
Spiraea, Honeysuckle, miniature Zinnias, Philadelphus,
Cornflowers, Marigolds, Lilac, Montbretia buds, and tiny
sprigs of wild Elder flowers. Leaves include Cupressus,
Rose, Acaena, Clematis montana, and Escallonia.

PLATE XIII ACCENT ON COLOUR I designed this plaque for a
television demonstration. The colours were chosen to
illustrate the striking colour effects that can be achieved
with preserved flowers. With a background of rich deep
blue I have used Dahlias, Marigolds, Helichrysums,
Montbretia, Spiraea, Daisies, Cornflowers and wild Elder,
with various leaves and ferns.

PLATE XIV ANTIQUE-FRAMED PICTURE An antique frame of deep dull blue, with gilt moulding, was used for this picture. Although badly chipped, the moulding was easily repaired with modelling clay and the frame repainted; see page 71. For the background of my crescent-shaped design I chose a rayon lining material in a shade of blue which toned with the frame. Pink, cerise and mauve flowers continued the antique theme. I used Hollyhocks, Dahlias, Carol Roses, Rambler Roses, Spray Carnations, Freesias, Heather, Lilac and Helichrysum, highlighted by creamy-yellow Narcissus, golden Cupressus foliage, and Choisya leaves.

Fig. 10 Flower arrangement plaque

[C] Using a separate piece of card make a smaller oval shape for the centre. Pad this with ¼ in. (6 mm) foam as for the flower arrangement plaque, and cover with fabric in the same way as described for oval plaques, page 48. Mark a central position for this on the colour oval shape and stick firmly in position. Choose plant material to tone with the colour of the card you are using and make your arrangement. An edging of smaller cut-out doyley motifs could be arranged and stuck directly around the padded shape as shown.

How to make the flower arrangement plaque (Fig. 10)

[A] Using a Stanley knife or similar sharp knife, cut a piece of thick cardboard to the required size. Round and cut the corners as shown, using a coin to draw round to ensure that all the corners are cut evenly. Cut a piece of ¼ in. (6 mm) plastic foam sheeting the same size as the carboard and glue together as shown.

[B] Mark and cut an oval letter box to the measurements shown.

[C] Cut a piece of felt and glue it over the foam. Clip the material inside the letter box as shown, pull it through to the back and stick it down with glue. Finish the outside edges and back of the plaque as described for square plaques on page 48.

Opposite above:
PLATE XV CLOCK 1

Opposite below:
PLATE XVI CLOCK 2

Above: PLATE 17 OVAL PLAQUE Helichrysums, French Marigolds and Hellebore were used for this design which is set against a background of chocolate-brown velvet. Glycerined Beech leaves and smaller plant leaves preserved by the desiccant method help to introduce contrasting shapes and forms. The spiky grasses which help to define the line of the design are the wild Quaking Grasses (Briza minor).

Opposite: PLATE 18 FLOWER ARRANGEMENT PLAQUE

[D] Using a suitably-sized flat-based top from an aerosol container, cut as shown. Glue section B to the back of your mount to form a container for your flowers and a support for standing. Glue a length of matching or contrasting cord round the edge, giving an attractive finish to your plaque.

[E] Fill the container with a piece of dry Oasis or plastic floral foam, and arrange your preserved flowers through the letter box.

Bases

The photographs of plaques illustrate my use of a variety of bases. Wood in all its many shapes, forms and finishes is ideal for plaques and swags which are to be composed mainly of seedheads and cones, etc. See Plates 3, 5 and 21. A light rub-over with a little linseed oil or polish will emphasize the grain of the wood and enrich the natural colouring of the wood. I find the small Surform tool illustrated in Plate 1 ideal for straightening and smoothing roughly sawn edges of wood; it can also be used to round off hard corners, and fine sandpaper can then be used to give the final smooth finish. A natural cane mat also makes a very suitable base for this type of plant material; see Colour Plate 1. Fig. 11 shows how to make covered bases for plaques such as those illustrated in Plates 17 and 18.

The four basic shapes used for plaque bases are square, oblong, round and oval. The round and oval shapes are much more attractive than square and oblong ones, unless you use some form of decorative edging.

Use stiff cardboard to prevent your plaque from warping (see details of types of cardboard for pictures, page 54). Polystyrene tiles cut to shape are also ideal.

Suitable fabrics
Felt is an ideal fabric for the beginner as it is easy to handle, does not fray, and can usually be obtained in a wide range of colours (see also page 40 for other types of materials).

Covering bases
Cut a piece of fabric slightly larger than your base but do make quite sure that you cut this on the straight grain of the fabric. Using a PVA or latex-type of glue, cover the entire base, spreading the glue quickly and evenly. Place your fabric over the glued base, smoothing it with your hand from the centre outwards to avoid bubbles. (See Fig. 11, parts 1A and B.)

Trim the excess fabric to within $\frac{3}{4}$ in. (2 cm) of the edge of your base (Fig. 11, parts 2A and B). Cut notches in the edge of the fabric covering of oval base, carefully turn these over on to the back of the base and glue down firmly and evenly (Fig. 11, part 3). With the square base it will only be necessary to clip and mitre the corners (as shown in Fig. 11, part 2A) as you turn the side edges over to glue down on the back of the base.

A simple and effective method of making a hanging for a plaque
A curtain ring may be attached by a small piece of tape which is firmly glued to the back of the plaque as shown (Fig. 11, parts 4A and B). Alternatively use a special sticky hanging tab which is available from art supply shops.

Neatening the back of your plaque
Cut a piece of paper slightly smaller than the size of your base and stick this over the back to cover the raw edges of the fabric and produce a neat finish.

Ideas for special occasions

Stately homes and other houses of interest are often decorated by flower arrangement societies to raise money for charity. It may be decided that a particular position such as a large expanse of wall needs a flower arrangement, but lacks a place on which to stand one. Providing there is a picture rail from which an invisible cord can be hung, swags such as the one in Plate 19 could be effectively used. Instructions for making the basic shape of this swag are given on page 49.

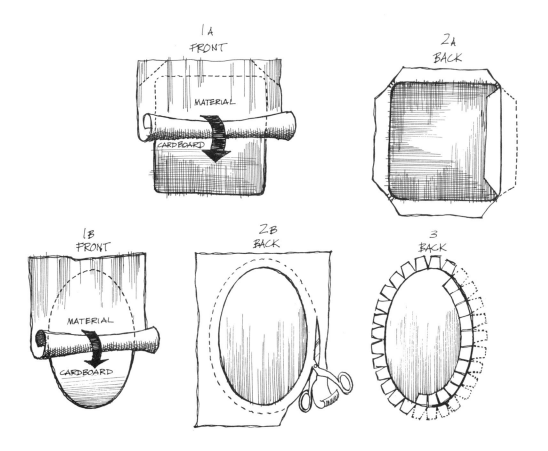

FRONT 1A

BACK 2A

MATERIAL

CARDBOARD

FRONT 1B

BACK 2B

BACK 3

MATERIAL

CARDBOARD

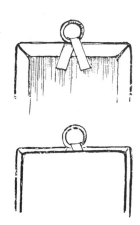

Fig. 11 Covering and hanging a square, oblong or oval plaque

How to make the basic shape for the flower festival swag (Fig. 12, Plate 19)

[A] Cut a thin strip of plywood not more than 2 in. (5 cm) wide and to the required length. Smooth with sandpaper, and paint a suitable colour.

[B] From a piece of stiff cardboard cut three oval shapes, graduating these in size as shown. From a block of dry Oasis (plastic foam) cut three oblong shapes, one to fit on each of the ovals, and glue them in position. Now make four holes, two on each side of the foam block as shown. Tie fine string or cord through them and knot securely at the back of each oval.

[C] Position each completed oval on the plywood strip as shown, and glue in position.

You are now ready to make your arrangements. Preserved flowers for this type of swag must of course have stems. If you want to preserve flowers with soft stems such as Anemones, you will have to wire them before preserving (see Fig. 2, page 26).

Above: PLATE 20 SWAG FOR A HEADBOARD Inspired by my
picture design in Colour Plate 10, this swag was designed to
decorate the headboard of the bed in the same room as my
swag shown in Plate 19. The flowers used were also in
toning colours. The principle of mounting the central and
corner designs was the same as for the large swag, each
being linked together by a double row of green cord. Tiny
floral motifs were attached to the cord at intervals.

Opposite: PLATE 19 FLOWER FESTIVAL SWAG (see also Fig.
12) This swag was used in the decorating of the Mayor of
Salisbury's house by our local floral arrangement society
for a flower festival in aid of charity. Measuring nearly 4 ft
(1·25 m) in length, this swag is one of a pair which I
designed and made to hang on the front panels of a white
wardrobe. Although the swag looks like three individual
swags it is in fact all mounted on one thin strip of wood
which was painted white to be inconspicuous.
The pinks and mauves of the flowers which include
Hollyhocks, Anemones, Dahlias, Freesia, Heather and
Statice were chosen to harmonize with the floral wallpaper
of the room. The necessary touch of lemon-yellow which
highlighted each arrangement was introduced by using
Narcissus and Polyanthus florets. The feathery foliage is
naturally-dried Cupresses with tiny leaves of Cyclamen
Neopolitanum and Tellima to fill in between the flowers. I
must emphasize that these swags were made for a floral
festival to give pleasure to visitors for a limited period.
Normally I would arrange flowers such as these under glass
in an airtight frame.

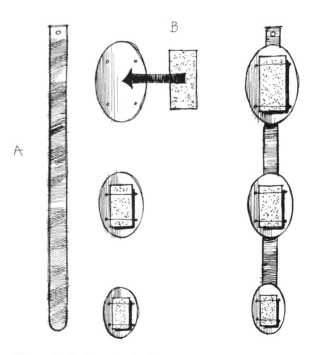

Fig. 12 Basic shape for the flower festival swag

Opposite: PLATE 21 NATURAL WOOD SCROLL This is an instance when one of nature's disasters can provide unusual and interesting material for the floral artist. This s-shape is made from a slice of diseased tree trunk. After chipping away the rotten wood from the centre I was left with only a small portion of hardwood which was then in the form of a letter C. This I sawed in half, reversed one half and glued it together again, making an ideal shape on which to create an interesting design. When using wood, remember that it is important to select and preserve plant materials which harmonize both in colour and texture. In the centre of this design I have featured three flower-like shapes which are the centre rosettes of Cedar cones glued on to short sticks. Other plant materials used include Cardoon Thistleheads, yellow Achillea, Old Man's Beard, and Poppy seedheads, together with Mahonia, Beech and Magnolia foliage. For fixing the plant materials which, with the exception of the cones, are all complete with their own stems, I used a large lump of Plasticine stuck to the centre. I bound it round with a piece of nylon stocking for additional anchorage, as shown for rolling pins in Fig. 22.

For a wood-panelled wall a swag of a similar design to the one shown in Plate 19 but made from cones and seedheads would be more in keeping, or alternatively a wall hanging such as the natural wood scroll shown in Plate 21. For a church harvest festival a plaque similar to the one shown in Plate 22 would create an interesting and unusual feature. *Note:* Chapter 5 should be read before attempting to make a plaque, as much of the information is both helpful and necessary for designing and making all types of plaques.

Right: PLATE 22 HARVEST PLAQUE An unusual idea for a church harvest festival. The plaited bread was made by the local baker. The base is made from stiff cardboard covered with creamy-beige linen-textured wall-covering. The bread shape was attached to this with glue. For the arrangement of autumn-coloured plant materials, Helichrysum, Moluccella, a miniature Bullrush, various wild grasses and glycerined foliage were used.

4 Designing and making picture frames and adapting standard picture frames

If you want to keep your preserved flower work in perfect condition for many years it is wise to make a picture rather than a plaque, particularly if you are using delicate flowers, as the glass will protect them from dust, dampness and, most important, inquisitive little fingers.

It is not as difficult as might be expected to frame three-dimensional flower pictures, and it need not be expensive. On the contrary, I designed and made many of the frames illustrated in this book from cardboard. Others I have adapted from ordinary picture frames, some of which I found in junk shops in poor condition and in need of restoration. This is certainly most rewarding, especially if the frame cost very little in the first place. Remember, a well chosen frame should enhance your picture. To do this successfully it must harmonize with both the plant materials and the background, and, most important, it should in no way dominate your picture. As you look closely at the illustrations you will see what I mean.

Picture frames from cardboard

In these days, with the accent on economy, it is not always possible to make as many preserved flower pictures as we would like. It is partly for this reason that I decided to devote a section of my book to designing and making attractive and inexpensive picture frames from cardboard. I also feel that it could provide a helpful stepping stone between making a plaque and making a picture using one of the standard types of frames describe on page 61.

Making frames in this way is not an entirely new idea. In Victorian times many picture frames were made from thick cardboard and covered with velvet. In fact, for the picture in Plate 23, I used just such a frame. I found it in a junk shop and after removing the picture I adapted it as described; see Fig. 13. The velvet covering of the frame was crumbling with age, so I replaced it with a piece of brown dralon left over from covering a chair. Although of rather heavy appearance, this frame is typical of the Victorian era.

There is of course no limit to the shapes that can be made from cardboard and so easily too as unlike wood they can be cut out using either a knife or scissors.

Types of cardboard

You will need two types of cardboard – a stiff one from which the rigid shapes are cut, and a more pliable one (the type that does not crack as you bend it) for side strips for frames such as those described in Figs 13 and 14. The stiff, rigid cardboard can often be obtained just for the asking, as it is used for making shop and office display cards. These cards are usually thrown away when no longer needed, and so it is worth making enquiries. Should this source of supply fail, you can buy thick mounting card from an art supply shop which will also stock the pliable type of cardboard.

How to make the Victorian velvet-covered frame (Fig. 13, Plate 23)
[A] Cut a cardboard shape of your own choice using stiff rigid cardboard. From the centre cut a shape for the flowers (keep this cut-out shape). For

PLATE 23 VICTORIAN FRAMED PICTURE Although re-
covered and adapted as shown in Fig. 13, the basic frame is
Victorian. This is an idea which could easily be adapted to
almost any shape that you care to think of. If I had made
this frame myself I think I would have cut the centre oval a
little larger. The arrangement of pink roses with their own
foliage and Antirrhinum florets with creamy white
Philadelphus, Deutzia, daisy-like Feverfew flowers, and a
few sprigs of wild Achillea is shown to advantage against a
background of chocolate-brown dralon. I am told that the
small shelves on each side of the frame were possibly used
to hold candles, but I felt my little cherubs were in keeping,
and so I wired one to each shelf. If you want to make
cardboard shelves, simply cover them and stick them on.

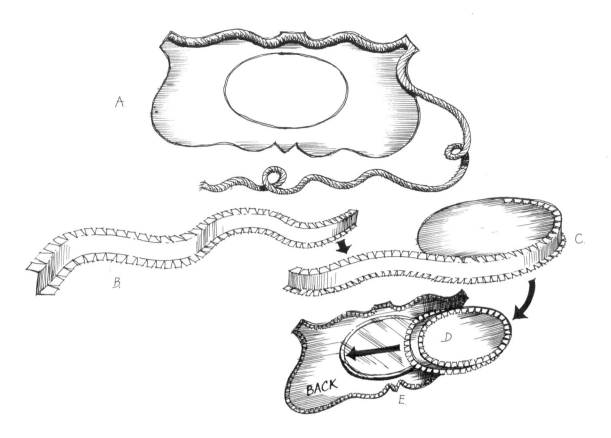

Fig. 13. Victorian velvet-covered frame

the edging use a length of very thick string or cord, sticking this in position with an impact glue such as Bostik No. 1. For covering, follow the details given for covering plaques on page 48.

[B] Using the cut-out shape for your mount, measure round the edge of it and cut a strip of pliable cardboard to this length. The width will depend on the depth you need to take your flowers without squashing them. Mine was about 2 in. (5 cm) plus 1 in. (2.5 cm) for fixing. Cut notches on each side of the strip to a depth of $\frac{1}{2}$ in. (12 mm). Bend one notched edge inwards (a) and the other outwards (b).

[C] Stick (a) to your mount as shown.

[D] Cover and finish this recessed shape as for round and oval pictures on page 65 and make your arrangement.

[E] To assemble the finished picture, cut glass, if required, slightly larger than the hole and glue it in position on the back of your frame. Glue notched edge of your recessed shape over the glass. Place a heavy weight on the back until glue is set.

How to make the frame for the triple Victorian-style picture (Fig. 14, Plate 24)

[A] Cut out your basic shape from a piece of stiff cardboard. Mark and cut out the three windows as shown.

[B] Take the measurement of the outside of the cardboard shape and cut a length of thick pliable cardboard to this size; the depth will depend on the size of flowers you are using. For the picture shown, in which I have used small flowers, $\frac{3}{4}$ in. (20 mm) is adequate, plus $\frac{1}{4}$ in. (6 mm) extra. From this extra, cut a notched edge for fixing.

[C] Glue all the tabs of the notched edge to the inside edge of the basic cardboard shape.

PLATE 24 TRIPLE VICTORIAN-STYLE PICTURE

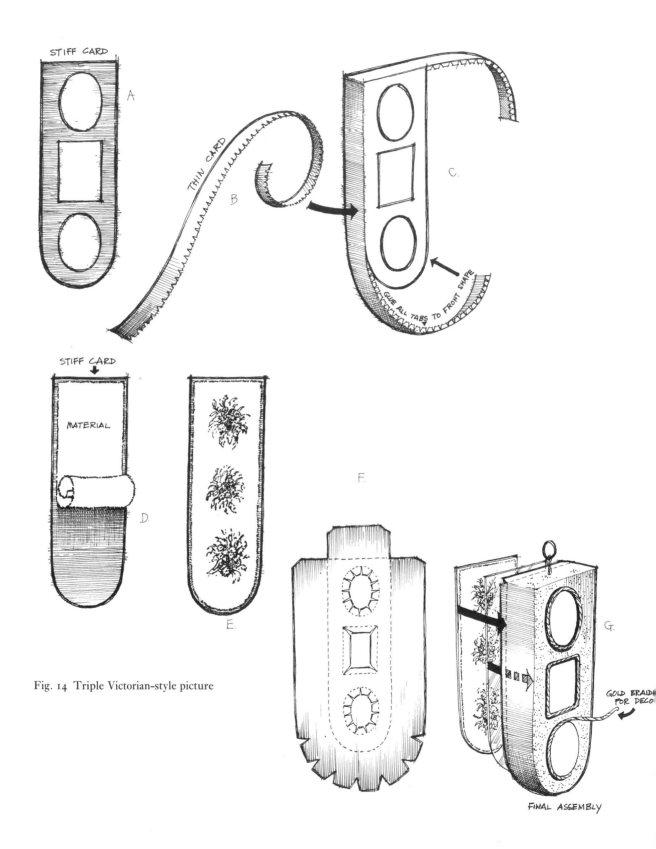

STIFF CARD

A

THIN CARD

B

C

GLUE ALL TABS TO FRONT SHAPE

STIFF CARD

MATERIAL

D

E

F

G

GOLD BRAID FOR DECO

FINAL ASSEMBLY

Fig. 14 Triple Victorian-style picture

PLATE 25 MINIATURE VICTORIAN-STYLE PICTURE Measuring
only 5 ×4 in. (12·5 × 10 cm), this picture is an adaptation of
the triple Victorian-style picture shown in Plate 24. The
frame is made in the same way: both are covered with
cotton furnishing velvet. Pictures as small as these require a
careful selection of tiny flowers; see Plate 10. Among the
flowers I used were Forget-Me-Nots, Miniature Roses,
Saxifrage, Daisies, tiny sprigs of Elder florets, Clematis
buds, and tiny leaves preserved by the desiccant method;
see Plate 12.

[D] Use the completed box shape and draw round it to obtain the exact size for the base of your picture. This should also be made of stiff cardboard, and covered with a piece of fabric cut slightly smaller than the cardboard.

[E] Using your box shape as a template, place this face downwards on the fabric-covered base and draw lightly round the inside of each window with chalk. Marking the shapes in this way will ensure correct positioning of each tiny arrangement.

[F] Cut a piece of fabric large enough to cover the top and sides of the box shape plus about $\frac{1}{2}$ in. (12 mm) extra all round for turning. Using a latex or PVA type of glue, such as Clam 7, stick the fabric in position over the top of the box shape and then mould and stick it in position round the sides. You will need to ease the fabric slightly round the curved part of the frame; with thicker fabrics you may have to make a few slight creases, but being underneath the finished picture, if carefully moulded they will be hardly noticeable. Leaving about $\frac{1}{4}$ in. (6 mm) of fabric extending beyond the cut-out windows, cut away remaining fabric. Now cut out notches as shown, turn the notched edges to the inside and stick down firmly.

[G] A stuck-on decorative cord makes a pretty edging to each tiny individual window. Pierce a hole in the top of the frame and through this thread a short matching length of cord with a small curtain ring attached. Knot the ends of the cord firmly inside the frame and cover the knot with a blob of glue for safety.

Have a piece of glass cut slightly smaller than the front of the cardboard shape. Secure this to the inside with glue. Squeeze glue round the un-covered edge of the base and stick this to the back of the cardboard box shape. Turn the edges of the material over this join, stick down and neaten at the back of the picture as described for plaques, Fig. 11, page 49.

Flower pictures for special occasions

It is fun to design and make a picture as a present to mark a special occasion such as a birthday or St Valentine's Day. I have illustrated two ideas of my own but these are only intended to stimulate the imagination. One example is the Valentine Picture, Colour Plate VI, instructions for which are given below. After seeing this at evening classes one of my students had the idea of making a picture for a friend's wedding and so she designed and created the heart-shaped picture in Plate 26.

How to make the Valentine picture frame (Fig. 15, Colour Plate VI)

[A] Find a suitable square box – the one I used was 10 × 10 × approx. 1 in. deep (25 × 25 × 2.5 cm).

[B] Mark and cut a heart shape in the centre of the lid as shown, but do be sure to allow enough space between the cutting line and the edge of the box for your chosen lace trimming.

[C] Cut a piece of fabric large enough to cover the top and sides of the box, plus about $\frac{1}{2}$ in. (12 mm) extra on the sides to turn in. Using a latex or PVA type of glue such as Clam 7, stick the fabric in position, cutting away the corners with a sharp pair of scissors as shown. Leaving about $\frac{1}{2}$ in. (12 mm) of fabric extending beyond the cut-out heart shape, cut away the remaining fabric. Now cut notches in the extended $\frac{1}{2}$ in. (12 mm) of fabric as shown, turn inside and finish as shown for the Oval Plaque in Fig. 11, page 49.

[D] Stick lace trimming round the top edges of the box. Measure round the edge of the heart shape, gather a length of lace to this size and glue the edge of the lace to the edge of the heart shape as shown.

[E] Measure and cut a piece of fabric to cover the inside of the box and glue in position.

[F] With the lid on the box, lightly draw round the heart shape to mark the position for your design. Remove the lid and arrange your flowers. Replace the lid and your picture is complete. If necessary use a little glue to keep the top and bottom of the box together.

Fig. 15 Valentine picture frame

Adapting standard picture frames

Standard picture frames cannot be used for preserved flower pictures unless they are adapted (by standard I mean the type of frame which is used for paintings, prints and photographs). This can be done quite easily by making a recessed box which will fit into the back of the frame.

PLATE 26 A WEDDING GIFT A picture designed and made as a wedding gift. The cardboard shape was made by the same method as described for round and oval recessed shapes for picture frames on page 65.

PLATE 27 ARRANGEMENT IN AN OLD PLASTIC CONTAINER
An ingenious idea from an eighty-six-year-old lady. The
centre has been cut from the lid of a plastic honey container
and replaced with a piece of plastic film. The container
itself has been cut down, leaving it just deep enough to take
the flowers.

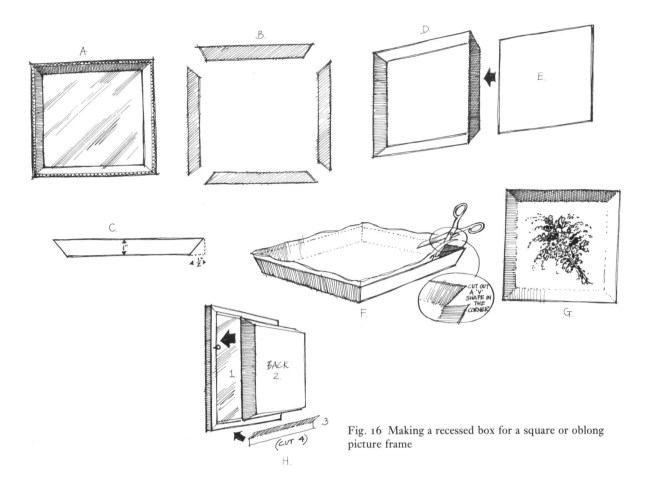

Fig. 16 Making a recessed box for a square or oblong picture frame

Making a recessed box for a square or oblong picture frame (Fig. 16)

[A] Use either an old frame cleaned and restored as described on page 71 or a new standard frame made for you by a picture framer. Make a recessed shape to fit in the rebate of your frame, as described below.

[B] Cut four strips of thick cardboard (or hardboard if preferred). Make sure these are wide enough to give the required depth to take your plant materials without squashing them. Take the measurement top and bottom on the inside of the rebate at the back of the frame, and cut two of the cardboard strips to this length. Now measure the two sides and cut the other two strips slightly short of this measurement by twice the thickness of the cardboard.

[C] Measure in $\frac{1}{2}$ in. (2.5 cm) on each end of one side and cut. Mark and cut the other three strips to correspond.

[D] Using a contact or woodworking glue, stick the four strips together with the two lengths which were cut short of the rebate measurement on the inside.

[E] Place this recessed shape in position in the frame (this will enable you to keep it rigid). Now mark and cut a piece of very thick cardboard (or hardboard) to fit the back. Glue these two sections together. Stronger adhesion will be obtained if you put a heavy weight on the back until the glue is set.

[F] To line the box, select a suitable piece of fabric, large enough to cover the base and sides of your recessed box. make quite sure that your fabric is crease-free before attempting to make a start, as

64

there is no way of removing the creases after the fabric is glued to the mount.

Cut your fabric on the straight, if necessary pulling a thread first to ensure accurate cutting. Make sure too that the fabric is kept straight while it is being glued; this will eliminate any chance of puckering. If the fabric has a prominent grain it is essential that for this reason alone it is kept straight because nothing can be more distracting to the eye than a fabric grain which is not running parallel to the frame.

Sticking your fabric can be rather a hazardous job until you get used to it, so I would advise the beginner to avoid very thin fabrics. Not only is it important to use the glue sparingly and to cover the entire surface of your mount to prevent bubbles appearing in the fabric, but it is important to use the correct glue. A PVA or latex-type is ideal, as opposed to the contact glue which is used to stick the plant materials. There are several suitable types on the market; Copydex is a particularly well known latex one, but personally I find this a bit thick and prefer to use a PVA one called Clam 7 which, if used as directed, will stick the finest of materials without seeping through.

Cover the entire base first, spreading the glue quickly and evenly. Position your fabric and with a clean hand smooth from the centre outwards. Now cover the sides one at a time in the same way, cutting the corners as shown. Use a sharp-pointed pair of scissors, make a clean cut and stick immediately to avoid fraying. I find that overlapping the fabric at the corners is too bulky. Wait until the glue is dry and then trim the fabric close to the edge of the frame all the way round.

[G] Assemble your chosen design; see page 76.

[H] Your picture is now complete and you need to fix frame and recessed box together. If glass is being used get a piece cut to size. If you are using an old frame the original glass can be reused if in good condition. (Check for distortion – common with old glass.) Wash and thoroughly clean glass, and then with your frame face downwards on the table rest the glass in the rebate. Position the recessed box over the glass. Cut four strips of thin pliable cardboard (cereal packet cardboard is ideal) the same length as the sides of the frame and an inch or so (about 3 cm) wide. Mitre the corners to correspond with the mitres of the frame, fold the strips as shown and stick firmly in position, fixing

one side of the fold to the back of the frame and the other to the side of the recessed box.

Mark position and screw in 'screw eyes' on each side of the frame. Then attach a cord for hanging.

Making a recessed box for a round or oval picture frame (including clocks)
(Fig. 17)

[A] Decide on a suitable picture or photograph frame, either an old one cleaned and restored (see page 71), or a new standard frame. Finding a picture framer who specializes in oval frames is not easy – see the List of Suppliers. In plate 28 and Colour Plate VII I used miniature plaster frames of the mass-produced type imported from Italy and usually sold with reproduction old Master prints in them. Plastic frames can also be found; these are often sold by florists, and although rather crudely gilded they can always be touched up with Gold Finger paint – see page 72.

Anyone with a lathe who specializes in wood turning could be approached about the possibility of making round frames, such as that shown in Plate 29.

[B] Measure round the inside of the rebate of the frame and cut a strip of thick cardboard to this size (the pliable type which bends without cracking). The width of this strip will depend on the plant materials you are using, but it must be wide enough to prevent them being squashed, plus an extra $\frac{1}{2}$ in. (12 mm). Cut notches as shown and bend over the extra portion. Cut a round or oval shape from thick cardboard (or hardboard) the same size as the rebate or picture area, less an allowance for the thickness of the cardboard strip. Glue the notched edge of the strip over the cardboard shape to form your box.

[C] To line the box, follow the basic instructions for square boxes, but with oval and round frames it will be necessary to cut the fabric in two pieces, i.e. one piece for the base and a separate strip for the side. Measure and cut these accurately. Assemble your chosen design; see page 76.

[D] For the final stage of your picture follow step H of the instruction for the recessed box for a

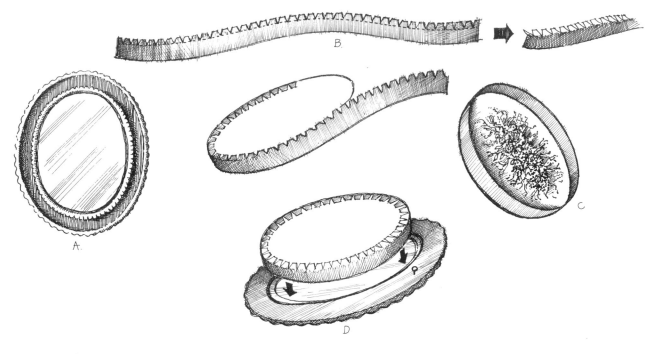

Fig. 17 Making a recessed box for a round or oval picture
frame

square or oblong frame, but instead of cutting four
individual strips of cardboard for fixing frame and
box together cut one single strip; this may need to
be notched to enable it to lie flat when bent round.
Mark position and screw in screw eyes on each side
of the frame. Attach a cord for hanging.

Victoriana (Plate 30)

For this picture I used a plain black ebonized oval
frame once used for a family photograph. I would
have preferred it with a beaded edge, and one day I
saw some strings of coloured wooden beads,
bought them, stuck them on the frame and then
painted the whole thing with Gold Finger paint. It
worked.

As gold card was used in Victorian times for
mounting pictures I thought I would try using it as
a background, but I could only obtain gold paper.
However, after sticking the paper to a piece of
cardboard I decided that the gold of the paper and
the gold of the frame fought with each other.

Thinking again of Victorian times I tried ruching
lace over the gold paper as an edging, but this
didn't look right either, and so I put it aside. Then
while browsing among secondhand stalls I found
some old lace mats with linen centres. I could
visualize the centres cut out and flower pictures
taking their place, if only one was the right size.
But they were both just too big. Not to be beaten, I
gathered the centre just enough to make it fit, and
so once again I became interested in making my
picture.

I mounted the lace on a separate piece of gold-
covered card, and then cut out the centre following
the shape of the lace. The finished mount I fixed
directly under the glass enabling it to be held in
place by the recessed box; see the previous pages
for details of a recessed box.

I chose cottage garden flowers in shades of pink
and blue to continue the Victorian theme, among
them Scabious, Lavender, Daisies, Lilies of the
Valley, Antirrhinums and Roses.

PLATE 28 MINIATURE OVAL GILT PLASTER
FRAME These attractive plaster frames, sold
with reproduction Old Master prints in them,
are cheap and easily adapted to make
charming miniature preserved flower pictures.
As they measure only $4\frac{1}{2} \times 4\frac{1}{2}$ in. (11 × 11 cm),
it is necessary to use tiny flowers to obtain the
correct scale of arrangement for the frame. I
have used individual florets of Rambler Rose
and Philadelphus including their buds, also
tiny lawn Daisies, sprigs of Forget-Me-Not
and a single floret of the shrubby Spirea.

PLATE 29 ROUND FRAMED PICTURE

Plate 30 Victoriana

Opposite: PLATE 31 INSPIRATION FROM A HOLLYHOCK

Another junk shop frame – this time only the paintwork needed a face lift. Longing to do a study in green and white to use my favourite white hollyhocks, I felt this frame would be ideal, being green on the outside with an inner moulding of creamy white. See page 61 for adapting frames.) A search through my oddment drawer revealed a piece of very old lime-green satin which was once the lining of an old velvet cape. It was in fact just the colouring of the Hollyhock centres, and so my picture was beginning to build up. Looking again into the centres of the hollyhocks I noticed the lime green shaded to a deeper green, and for a contrasting form of plant material I decided to use green feathery fern fronds. Some of the hollyhocks' centres showed a hint of gold and so I introduced some of the gold everlasting flowers called Lonas. The sprays of tiny white flowers are shrubby Spirea, used with a few Quaking Grasses to complete my design.

Fig. 18 Repairing picture mouldings

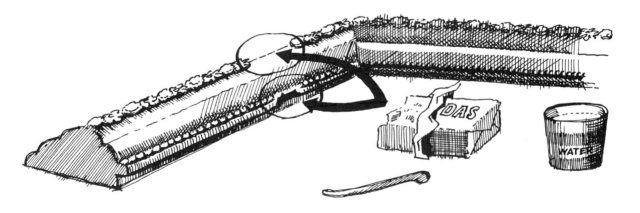

Restoring old or antique frames

Moulded plaster frames

See Plate 31 and Colour Plate VII. These frames are usually rather ornate, and particularly susceptible to damage unless carefully handled. For this reason, when you find this type of frame in a junk shop it will usually have at least a few bits of moulding missing. Replacing them is quite an easy and rewarding job for the novice to tackle (see Fig. 18), but frames which have a large amount of moulding missing should be only dealt with by an expert or someone with previous experience.

Cleaning the frame

First remove the existing picture (if it has one) and the glass, if there is any. If the glass is in good condition, it can be cleaned and re-used to save expense. Brush off all the loose dirt. Avoid using water on plaster frames, as this can cause the plaster to disintegrate. Instead, wipe or brush them over with a little methylated spirit.

Restoring the paintwork

Quite often the paint is restored to its original colour after cleaning, but if pieces of moulding have had to be replaced, these will need to be touched up with a matching paint. If, however, the paint seems rather faded, it is a good idea to give the complete frame a facelift with a light brushing of paint.

I have experimented with different finishes for old frames in an attempt to regain their original charm. It is easy to spray frames, but this tends to give them rather a crude finish which does not harmonize well with preserved flowers. You will only need small quantities of paint. The small tins sold for model making are quite good and come in a wide range of colours, but most of them are only available in a gloss finish which gives a rather 'painted' look. Recently I started experimenting with artists' oil paints: the little tubes are inexpensive and from a small range of basic colours it is possible to mix the exact shade for almost any frame. Thin oil paint with a little white spirit and brush it on sparingly. This gives a very good matt finish, with a restored rather than a painted appearance, but unlike ordinary paint it will take several days to dry completely. If you prefer a slight gloss effect, give the frame a light spray with an aerosol varnish when dry.

Retouching gilt frames

For gilt frames or gilt inserts in frames I find Gold Finger excellent. This is sold in tubes in the form of paste, and should be applied very sparingly with a brush or your finger. Alternatively a little white spirit can be mixed with the paste which can then be applied very thinly with a brush; this is very suitable if the gilt is in good condition but just in need of a facelift. Gold Finger is made by Rowneys and comes in several shades. I find that Antique Gold matches the gilding on most old frames, but sometimes I need to mix this with a little Copper Gold Finger to give the warm gold colouring which antique frames often call for.

A tube of Gold Finger will last for a very long time without deterioration. There are other brands of this type of gold paste on the market which produce the same effect, but at the time of writing I have not tried them. Most of the gold paints which are sold in liquid form I find give a too new-looking finish.

Repairing mouldings (Fig. 18)

You will need the kind of modelling clay that hardens naturally, without firing. Several brands are available and can be purchased at most art and craft shops. I use 'Das' which is also available at large stationers. This is moulded directly on to the frame with water, using either your fingers, a small spatula, or even a cocktail stick depending which you find the easiest to manipulate. I often use a plastic ice cream spoon (using both ends). Follow the directions for use as supplied with the clay, and build up the missing sections by modelling them into shape to match the existing moulding.

If your first attempt is unsuccessful, this type of clay can easily be removed by softening with water.

5 Designing and making a picture

Pictures need to be more than just a collection of shapes, textures and colours. You must blend plant materials to form a design or arrangement which is pleasing to the eye.

Inspiration for designs can come from many sources: wallpapers, fabrics, carvings, and of course books on design and interiors. As a beginner, start with a relatively simple design; these designs often prove to be the most effective. Various aspects of a picture or plaque will govern your choice of design: the size and shape of your frame or mount (as you will see in Fig. 19, some designs adapt better than others to a particular shape).

Fig. 19 Adaptation of design to shape of frame or mount

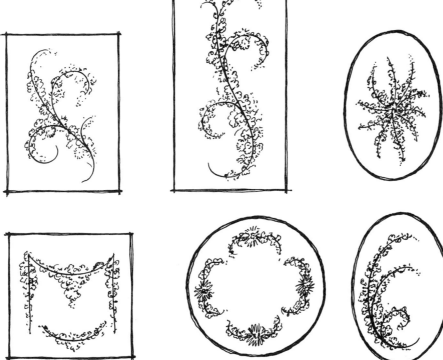

Opposite: PLATE 32 JAPANESE-STYLE FUCHSIA
PICTURE Simple yet effective. These twisted silver-grey lichen-covered branches of Wild Heather with cerise-coloured Fuchsias gracefully hanging from their tips are arranged to resemble one complete flowering branch. In this design the spaces created by the branches are as important to the success of the picture as the branches themselves. As a background material I chose black cotton velvet which forms a striking colour contrast to the plant materials. The frame is an old wooden one which I painted with a mixture of silver and black paint to harmonize with the picture; see page 64 for adapting frame.

Fig. 20 Space in relation to your design

Space

Space in relationship to your design is important. By this I mean the amount of space surrounding your design, which helps to create a pleasing visual balance within the frame. Look at Fig. 20. The trained eye will immediately see that the drawing above has too much space for the size of the design, or in other words, the design is too small, while in the one below the amount of space is just right, and creates a pleasing balance. It is of course the design and its relationship to the frame which creates this space. Take away the frame, and we then begin looking at space within the design (see Plate 32). Here the spaces are just as important as the branches; between each branch is a pleasing shape. Even the design in Plate 2 has pleasing spaces between the outline. Some designs (for example, Plate 29) are composed of several smaller designs.

Regardless of the type of design or arrangement, spiky plant materials are generally used to form the outline, while the rounded more solid forms are kept towards the centre, where they create a better balance and form the focal point of the design. Whatever the order in which you collect the necessary ingredients for a picture, i.e. plant materials, background material and frame, it is important to have them all at hand before attempting to begin making your picture.

A good design or arrangement is something which can only be achieved with practice. Try to be your own critic, and with each completed picture ask yourself how you could have improved it. Study the simplified drawings of designs in Fig. 19 which are the basis of many of my pictures; beginners designing their first picture will find them a useful guide.

To copy exactly a design or arrangement from a picture is almost impossible, for it is most unlikely that you will have plant materials of exactly the same shape and size.

Method of working

The frame
First take your frame (if you are using a standard picture frame) and make a recessed shape for it (see page 64 for square and oblong frames, and page 65 for round and oval ones). See also pages 54–60 for home-made cardboard frames, or should you only be making a plaque, see page 48 for details of preparing mounts.

The design
Taking into consideration the shape of your frame, decide on the basic design you are going to work from. Your design can then be created in one of three ways:

1. Draw your design on a piece of plain cardboard and use this to practise arranging your plant materials. Once you have achieved something which pleases you, transfer your materials piece by piece into your lined recessed box.
2. Make a few guiding lines with a piece of chalk directly on to the background material, and work from this.
3. Create a completely spontaneous design working directly with the plant materials, which is the way I personally prefer to work.

Assembling the plant materials
Regardless of your chosen design the method of assembling and sticking your plant materials is basically the same, with the exception of the pictures shown in Plates 18, 19 and 21. Here I have used the plant materials in the same way as for an ordinary flower arrangement, complete with either their own stems or false ones, and arranged in plasticine or Oasis.

I always like to assemble the main part of my design before I begin glueing. First I place a few spiky plant materials which will give me the basic outline and flow of my design. I then begin to build up my design by creating the focal point or centre of interest, usually consisting of the larger, more important and denser plant materials. Remember to use only plant materials which are in scale with the size of your picture: for example in Colour Plate VIII I have used Dahlias towards the centre of my design but in the miniature picture shown in Plate 28 I have used Miniature Roses.

In traditional designs, you must create a link between the outside spiky plant materials and the larger central flowers. This is achieved by using plant materials which are not as spiky as those on the outside and not as solid as the ones used in the centre. By doing this the eye will be led gently through the arrangement or design, avoiding the sudden jump from solid to spiky materials. A sense of depth in your picture can be created by overlapping plant materials.

Creating a natural flow or sense of direction in your design
You will notice that in most of my pictures flowers are placed slightly on their side. Arranging them in this way helps to create a natural flow and also to determine the intended direction of the design.

Padding the centre of the design
Occasionally I find that, in spite of using larger flowers, the centre of my design still appears rather flat. This problem can be overcome by using damaged or unwanted flowers underneath to act as a padding. These must of course be placed in such a way that in the finished picture they are completely hidden.

Sticking the plant materials
Latex-type glues are easiest to handle but I find that they take too long to set, which means that your plant materials cannot easily be fixed at the required angle; this in turn tends to create the flat effect we are trying to avoid. The ideal glue is one which enables you to place a flower in any position and dries fairly firmly after a few seconds, holding the flower. This means using one of the contact glues, such as Bostik No. 1 or Evostik, but it is important to squeeze out only a very little at any one time, otherwise the glue will dry on the tube before you use it all.

Working from the tube, squeeze a spot of glue directly on to the back of the plant material. The exact position will of course depend on the angle at which you wish to stick them. See Fig. 21.

Once your design is complete you are ready to carry out the final stage of your picture. However, I prefer to leave my picture until the next day. I then give it a final critical inspection, for it is at this stage, having come back to it fresh, that I can often

Fig. 21 Putting the picture together

see the need to add another leaf or flower, or maybe see something that requires slight alteration or adjustment.

Cleaning the picture

This is an important step in making a preserved flower picture and should not be overlooked. Turn your picture face downwards and tap the back quite hard all over. This will remove any loose bits, and of course should there be a leaf or flower not securely stuck it is better to find out at this stage, rather than later. To ensure there are no bits still clinging to the background of the picture I carefully brush the open spaces of the background material around the plant materials, using a small artist's oil paintbrush. This is especially necessary with dark-coloured velvets which show every speck of dust. Do take care not to damage any of the plant materials.

Final assembly of your picture

For fixing glass, frame and recessed shape together, see the directions given with the particular type and shape of picture you are making (page 64 for oblong- or square-framed pictures, page 65 for round or oval ones and page 54 for cardboard-framed pictures).

Displaying your picture

Having created a picture of which you are justly proud, it would be very sad to find that after a very short time the plant materials had lost all their colour. This can happen, of course, but only if your picture is exposed to extremely strong light, heat or damp. Avoid this by observing the following points when choosing a position to hang your picture:

1. Hang it away from direct daylight – this applies to light from any direction: north, south, east or west.

2. Hang it away from the direct heat of a fire or radiator.

3. Avoid hanging it in a room that is known to be damp.

6 Novelty pictures and wall hangings

There are a surprising number of things which can be used as frames for preserved flower pictures, and as bases on which decorative wall hangings can be made.

In the following pages I have presented a few of my ideas. I feel they are novel rather than gimmicky, and I hope there is something for every home.

Clock 1 (Colour Plate XV)

I confess to being a great hoarder of almost anything which I feel may come in useful, and when I was offered an old broken clock I thought I might well be able to use it as the basis for a picture.

Details for conversion
I removed the works and back from the clock, leaving only the case and convex glass. The case was covered with a dark shiny varnish, which I removed with paint stripper to reveal the lovely colouring of the natural wood. A little brown shoe polish rubbed well into the wood gave a natural-looking finish, and also emphasized the wood graining.

The arrangement
Fig. 17 shows how to make a box shape to take the arrangement, which is composed of tiny Larch and Alder cones, with orange miniature Zinnias, individual florets of Monbretia and Ammobium. The muted colouring of the cones and flowers complement the wood of the clock case, and are shown to their best advantage against a background of cream linen.

Many otherwise useless clocks could be used in this way, providing the works can easily be removed.

Clock 2 (Colour Plate XVI)

This old gilt alarm clock is quite different from the wooden one. Being gilt, it calls for a completely different choice of plant materials, and so I have used Roses, Forget-Me-Nots, individual florets of white Lilac, Syringa and shrubby Spirea with tiny Rambler Rose leaves and ferns. For the background I chose dark green velvet.

The box was made and lined in the same way as for the first clock.

Grandfather's watch (Plate 33)

Converting old clocks into flower pictures led me to think of old watches. This silver one was mechanically useless, so I removed the works and fixed a small circle of dark blue velvet in the back of the case. I chose miniature pink Roses, Forget-Me-Nots and white Saxifrage florets, with tiny sprigs of Cupressus foliage to create a miniature design in the correct scale for the watch.

Curtain ring hanging (Plate 34)

This brass curtain ring with a circle of velvet or felt glued to the back makes an ideal frame for a simple, cheap miniature wall hanging. My choice

Plate 33 Grandfather's Watch

PLATE 34 CURTAIN RING HANGING

PLATE 35 ROLLING PIN 1

of background was dark green velvet, and my flowers include Miniature Roses with a bud, an individual floret from a Delphinium, Forget-Me-Nots, May blossom, and tiny fronds of preserved fern.

Rolling pin 1 (Plate 35)

Rolling pins hold a particular fascination for me because of their beautiful wood graining. I also feel that, unlike many other types of pictures and plaques, they make ideal wall hangings for the kitchen.

The rolling pin in Plate 35 is a particularly fine example of wood grain. A light rub over with clear polish is all that is necessary to give the wood a satin finish, but first fix the Plasticine (Fig. 22) or you may find difficulty in getting it to adhere.

I chose my plant materials carefully to go with the rich honey colouring of the wood. Glycerined leaves of Copper Beech and Solomon's Seal, yellow Achillea, golden Wheat, seedheads of one of the summer-flowing Lilies and Fir cones (see Fig. 1 for wiring these). The three flower-like shapes in the centre are in fact Fir cones cut in half. Notice the attractive pattern on the inside of the cones which is revealed when they are cut.

Fasten a piece of fine cord to the handles to hang the rolling pin.

Rolling pins – how to hold the plant materials (Fig. 22)

For holding the arrangement of flowers, I use a lump of Plasticine moulded and pressed firmly in position, with a piece of old nylon stocking tied round for additional support.

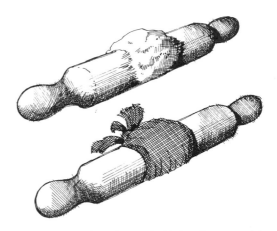

Fig. 22 Rolling pins – how to hold the plant materials

Rolling pin 2 (Colour Plate II)

Made from a different type of wood, this rolling pin has a less pronounced grain and in colour is a soft greyish mauve. Just as the other rolling pin inspired me in my choice of plant materials, so did this one. If you compare the two you will see how the texture and colouring of the wood has influenced me. For this rolling pin I used steel-blue Echinops, Wild Heather and cerise Helichrysum with Poppy seedheads, fluffy cultivated Mare's Tail Grasses, and trailing wild Barren Brome Grasses which have retained their beautiful mauve colouring.

Honey-coloured wooden spoon (Plate 36)

Tiny Fir cones, Alder cones, pressed Ferns, glycerined Beech leaves, Cupressus foliage and Quaking Grasses in shades of brown, cream and green blend with the rich honey colour of the wood. The centre 'flower' is in fact an orange Cape Gooseberry seed capsule cut open to form a petal-like shape.

PLATE 36 HONEY-COLOURED WOODEN SPOON

PLATE 37 MATCHING FORK AND SPOON

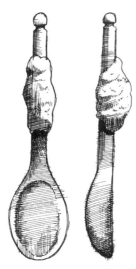

Fig. 23 Wooden spoons – how to hold the plant materials

Matching fork and spoon
(Plate 37)

In contrast, the dark mahogany wood of this fork and spoon lends itself to the arrangements of mixed flowers in shades of pink and blue with wild grasses and a few sprigs of glycerined leaves.

Wooden spoons – how to hold the plant materials (Fig. 23)
With these it is usually only necessary to mould and press a lump of Plasticine to the handles, as the plant materials are so much lighter in weight than the ones needed for rolling pins.

Postscript
This book has been based on my own collection of ideas. If in writing it I have inspired my readers and helped them to create their own original preserved flower pictures then I shall have fulfilled my aim.

List of suppliers

Decorative grass seeds
Thompson & Morgan, Ipswich, Suffolk, supply named varieties.

Maureen Foster's flower preserving crystals
M/F Crystals (Dept B3), 30 The Avenue, Woodland Park, Prestatyn, Clwyd LL19 R9E.

Oval picture frames in a range of types and sizes, to order from:
F. J. Harris & Son Ltd, 13 Green Street, Bath, Avon.

Index

87